CW00725252

PLANT NUTRIENT DISORDERS 5
Ornamental Plants and Shrubs

PLANT NUTRIENT DISORDERS 5

Ornamental Plants and Shrubs

G.C. Cresswell
R.G. Weir

NSW Agriculture

Inkata Press
Melbourne • Sydney • Singapore

INKATA PRESS
A DIVISION OF BUTTERWORTH-HEINEMANN

AUSTRALIA	BUTTERWORTH-HEINEMANN	22 Salmon Street, Port Melbourne, Victoria 3207
SINGAPORE	REED ACADEMIC ASIA	
UNITED KINGDOM	BUTTERWORTH-HEINEMANN Ltd	Oxford
USA	BUTTERWORTH-HEINEMANN	Newton

National Library of Australia Cataloguing-in-Publication entry

Cresswell, G. C. (Geoffrey Charles), 1953- .
Ornamental Plants and Shrubs

Bibliography.
Includes index.
ISBN 0 909605 88 2 (series).
ISBN 0 909605 93 9 (v. 5).

1. Nutritionally induced diseases in plants. 2. Plants, Ornamental – Diseases and pests. 3. Plants, Ornamental – Nutrition. 4. Ornamental shrubs – Diseases and pests. 5. Ornamental shrubs – Nutrition. I. Weir, R. G. II. Title. (Series: Plant nutrient disorders; 5).

635.923

Published by Reed International Books Australia Pty Ltd trading as Inkata Press

Edited by Patricia Sellar

Designed by John van Loon

Typeset in 10pt Goudy Old Style by Linographic Services Pte Ltd

Printed by KHL Printing Co Pte Ltd, Singapore

CONTENTS

Acknowledgments

The authors wish to thank all those people who have helped in some way with this book series. Although most of the photographs featured in the series are from the authors' own collections, some of the more unusual material was generously provided by colleagues. In this final volume, photographs were obtained from Prof. Dr. D. Alt, Osnabrük, Germany; Keith Bodman, Queensland Department of Primary Industry; Kevin Handreck, formerly CSIRO Adelaide; and Dr Alex Wissemeier. Most of the studio photographs were taken by Lowan Turton NSW Agriculture now at EMAI, Camden.

The manuscripts were typed by Margaret Daniels and Susan Mifsud and their patience and skill is much appreciated. The series has been expertly edited by Patricia Sellar.

The assistance of NSW Agriculture's advisory staff in obtaining field specimens and background information to help with a diagnosis is gratefully acknowledged. In the area of ornamental horticulture, special thanks is due to Liz Maddock, Greg Ireland and Betina Golnow. Over the years, generous help and encouragement has also been provided by colleagues at Rydalmere. We have also appreciated our contacts with growers, consultants and others in industry especially Graham Price and John Glendining.

PREFACE

This manual is part of a series called Plant Nutrient Disorders that aims to help farmers, advisers and students to identify nutrient deficiencies and toxicities in a wide range of crops and pasture plants.

In all, there are five manuals, dealing with the identification of nutritional disorders and they are:

1 Temperate and Subtropical Fruit and Nut Crops
2 Tropical Fruit and Nut Crops
3 Vegetable Crops
4 Pastures and Field Crops
5 Ornamental Plants and Shrubs

Each book describes and shows, with the aid of coloured plates, typical symptoms of the common nutrient deficiencies and toxicities found in commercially cultivated plants. Ways of distinguishing symptoms caused by a nutrient deficiency, toxicity, or some other problem such as a disease, herbicide damage or moisture stress, are explained.

These methods are a distillation of the experiences of nutrition chemists, advisory horticulturists and agronomists from the New South Wales Department of Agriculture who collaborated for more than thirty years to provide a diagnostic service to the farming community. This consultancy service ceased in 1988, but it is hoped that the skills developed for diagnosing nutrient disorders in plants will be passed on through each of these books.

Most of the coloured plates used in the manuals are of field specimens and, in many cases, leaf analysis has been used to identify the problem. The photographic and analytical records which are the primary sources of material in these books go back to 1958 and represent one of the oldest and most crop diverse collections of its type in the world.

This series provides leaf analysis and sampling methods for the most commonly grown crops in Australia, as well as for some poorly documented species. Where months are mentioned they refer only to the southern hemisphere. For northern hemisphere readers, the corresponding time of season or stage of crop development is shown in brackets. The booklets explain in everyday language the main uses for leaf analysis, how to interpret a leaf analysis report and the main shortcomings of the procedure. The information contained in this series should be helpful to anyone who wishes to understand crop nutrition better and build reliable diagnostic skills.

CHAPTER 1

PLANT NUTRIENT NEEDS

Sixteen elements are known to be essential for the healthy growth of ornamental plants. Three of these – carbon, hydrogen and oxygen – are obtained from the air by leaves and from water by roots. The remainder – nitrogen, phosphorus, potassium, sulphur, calcium, magnesium, iron, manganese, copper, zinc, boron, molybdenum, and chlorine – are primarily obtained by plant roots from the soil solution although small amounts may be absorbed through bark and leaves. Other elements, such as silicon, sodium and cobalt, are beneficial under some circumstances but are not used routinely in the commercial production of horticultural crops.

Nitrogen, phosphorus, potassium, sulphur, calcium and magnesium are needed by plants in relatively large amounts and are, therefore, known as 'major' or 'macro' nutrients. They normally accumulate to percentage levels in plant tissue. The remaining nutrients are needed in much smaller amounts and are called 'minor' or 'trace' elements. They are normally found at ppm (parts per million) levels in tissues. In addition to these nutrients, plants accumulate many non-essential elements in quantities largely determined by their availability in the growing medium.

An element is considered to be 'essential' when plants cannot grow and/or reproduce without it. Thus, when there is too little of an essential element, plant health suffers. Problems also arise when an element (essential or non-essential) is oversupplied. When plant performance is inhibited by the nutritional condition of the soil or root medium, the plant is said to be suffering a nutrient disorder.

There are two types of nutrient disorders:

- **Deficiencies** where the supply of an essential element is not adequate to sustain optimum plant performance.
- **Toxicities** where the supply of an essential or non-essential element is more than the plant can tolerate.

Figure 1 shows how the supply of an essential element affects crop performance.

Causes of nutrient deficiency

The two main reasons for a deficiency are that the amount of a nutrient in the growing medium is low or that it is not in a form available for plant uptake.

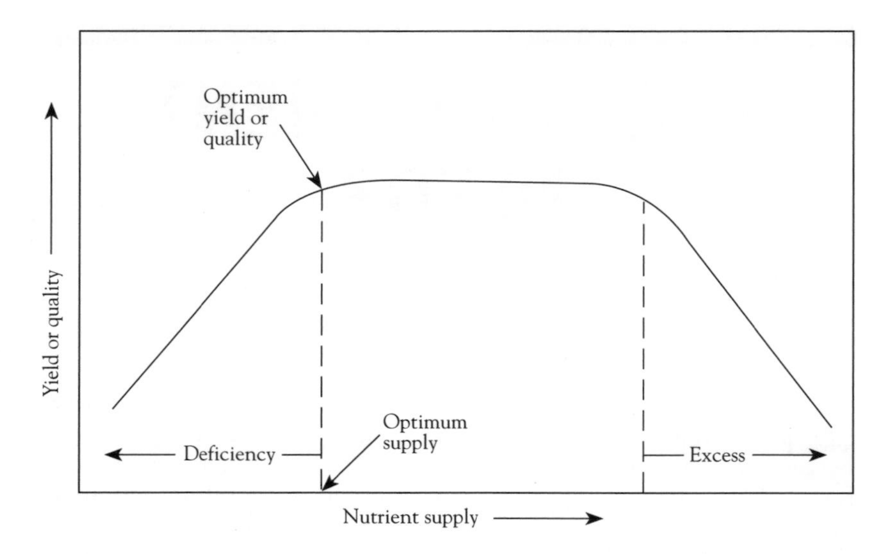

Figure 1 The relationship between crop yield or quality and nutrient supply

- **A low total level of nutrient in the growing medium**

Soils Nutrients which are readily water soluble can be depleted below levels which will support strong plant growth by leaching and erosion, processes which are accelerated by cultivation and clearing of land. Significant quantities of soil nutrients can also be removed in produce and crop residues, and fertility is reduced if these losses are not replaced by fertilisers or manures. Some soils, especially highly weathered soils derived from sedimentary rocks like sandstones and shales, have inherently low nutrient reserves and often have little capacity to retain applied nutrients against leaching. The most fertile soils are formed from weathered igneous rocks rich in minerals, or from mineral-rich material deposited on river flood plains. Fertile soils generally have a high content of clay and organic matter which provide sites for cation exchange and adsorption of nutrients.

Soilless media Although commonly referred to as soil mixes, modern growing media used with seedlings and container plants, and increasingly for cut flower production and landscaping, normally contain little if any soil. These soilless media consist mainly of an organic component such as peat, bark or sawdust together with an inorganic material like sand, vermiculite, perlite or polystyrene. In general, these ingredients have little natural fertility and only a modest cation exchange capacity compared to soil. Accordingly, crops grown in soilless media must be fertilised regularly and sparingly to prevent deficiencies or toxicities and to minimise losses of nutrients through leaching.

The low cation exchange (nutrient retention) properties of soilless media make it easier to alter the nutrient conditions around plant roots. However, with this improved capacity for the manipulation of crop nutrition comes an increased opportunity for disorders. Plants in soilless media, especially those based on inert materials like rock wool, perlite, vermiculite or sand, must be supplied with most of the essential elements in the fertiliser program and consequently can suffer deficiencies and toxicities of any one of them. By comparison, plants in soil or soil based media generally only need nitrogen, phosphorus and potassium, and perhaps calcium or magnesium, because the other nutrients are normally present in sufficient quantities. Organic materials like bark, sawdust, peat and coir dust do contribute some nutrients but supplies are usually poorly balanced and inadequate for optimum plant growth. The supply of nutrients to plants in containers, particularly cell grown seedlings, is also restricted by the small volume of medium available to the roots.

- **The nutrient is not present in a readily available form**

Plants obtain most nutrients from the soil solution and so nutrient sources with low solubility in water, such as some mineral phosphates or coarse grades of lime and dolomite, are not readily available for uptake. The availability of applied nutrients is influenced by chemical reactions in the growing medium which affect their solubility. For example, aluminium and iron, often very plentiful in acid soils, can form insoluble compounds with phosphorus. This phosphorus 'fixation' is less important in soilless media which generally have lower aluminium and iron contents. High concentrations of one element in the soil solution can reduce the availability of another nutrient. For example, phosphorus forms insoluble salts with calcium and the trace elements iron and zinc. Some trace elements including copper and manganese can form organic complexes which reduce their availability.

The acidity or alkalinity of the soil solution (measured by its pH) has a major influence on the availability of nutrients for plants. Nutritional disorders are more common when the medium is either alkaline or very acid, rather than when the pH is neutral (pH 7) or only slightly acid (pH 6). The pH of soilless media can drift from the optimum range relatively easily because of poor buffering. Iron, manganese, zinc and boron deficiencies are likely when the medium becomes too alkaline, perhaps as a result of over-liming. A drift to acidity, as happens when acidifying fertilisers such as sulphate of ammonia are used, can lead to deficiencies of calcium, magnesium, nitrogen, phosphorus or molybdenum.

The activity of soil micro-organisms during the breakdown of organic matter can influence nutrient availability sufficiently to restrict plant growth. Deficiencies develop when these organisms compete with the crop for some limiting nutrient, or toxicities may result when their activity releases nutrients for root uptake. Use of incompletely composted materials in a potting mix, or mixing plant residues through a soil which are high in carbon (C:N ratio >10:1), for examples, sawdust or straw, creates a strong biological demand for both nitrogen and phosphorus which can lead to deficiencies. This effect on nitrogen availability is referred to as 'nitrogen drawdown'. The availability of trace elements, such as manganese and iron, is also linked to biological activity. Some sewage sludges and manure-based composts have been found to bind manganese and copper so tightly in biological forms that sensitive plants like palms become deficient. By contrast, toxic amounts of manganese can be released through the changed biological conditions which follow waterlogging or pasteurisation of organic media or soil.

Causes of nutrient toxicity

Common causes of nutrient toxicities in ornamental crops include the overzealous or inappropriate use of fertilisers and trace elements, rapid nutrient release from controlled release fertilisers (CRFs) triggered by high soil temperatures, enhanced availability of certain nutrients under acid or anaerobic growing conditions, and the use of media or water containing high levels of salt or heavy metals.

When excessive amounts of fertiliser have been used, plant injury is commonly the result of nitrate or ammonium toxicity. However, some plants, particularly in the Proteaceae family, are very sensitive to phosphorus toxicity. Ammonium toxicity is often associated with the heavy use of animal manures and urea or with ammonium fertilisers, such as sulphate of ammonia and monoammonium phosphate (MAP), in cold wet soils. Banding of concentrated soluble fertilisers too close to the roots is a common cause of injury to young plants and germinating seeds.

High concentrations of almost any element in the media solution can lead to toxicity. Boron, copper and zinc have narrow optimum ranges (between enough and too much) and need to be used carefully. Many trace element toxicities arise when these elements are applied to correct or prevent a deficiency, either because excessive amounts are used or because the product is unevenly applied to the medium. When applying borax to soils, for example, any lumps should be crushed and the powder spread evenly over the treated area to avoid pockets of high boron concentration. Boron is best applied as a soil drench to crops in soilless media, in the fertigation program, or as a foliar spray. However, trace element sprays can cause injury if product recommendations are not followed. Soluble copper salts are particularly phytotoxic and can burn soft-leaved foliage plants. The risk of injury following a foliar spray can be reduced by 'neutralising' the copper solution with hydrated lime, or using a low soluble copper source such as copper oxychloride.

Trace element toxicities occasionally develop in crops which have been grown in contaminated media or which have been irrigated with contaminated water. Sawdust derived from softwood timber treated for pests can contain toxic amounts of boron or copper; rice hulls, some eucalypt barks and sawdusts can be high in manganese; and sewage, municipal and mining wastes and furnace ash are often contaminated with a variety of heavy metals including zinc and copper. Water or fertiliser solutions are sometimes contaminated with zinc or copper from corroding galvanised iron or copper pipes or brass fittings. High iron in irrigation water can stain or even damage plant foliage and flowers. The small quantities of fluoride in some groundwaters and in low grade superphosphate can injure sensitive plants like petunia, chordyline, gladiolus and dracaena in soilless media, but it is usually not a problem in soils due to chemical fixation. Concentrations of fluoride in drinking water are generally too low (around 1 ppm) to cause toxicity through root uptake but can shorten the vase life of some flowers including freesia, gerbera, gladiolus and chrysanthemum when absorbed through the cut stem.

Chloride, sodium and boron toxicities are the most common disorders associated with saline irrigation water or salty soils. Naturally saline soils are usually found in arid and semi-arid areas where rainfall is too low to leach salt from the soil surface. Clearing native timber and irrigating poorly drained land encourages salinisation of agricultural land by raising the watertable bringing salt into the root zone from deeper within the soil profile. Salts are often concentrated to very high levels in low or seepage areas, above impervious layers in the subsoil, or along the edges of furrows by leaching then evaporation. Sufficient salt can be absorbed through the leaves from sea spray or from saline irrigation water to injure sensitive plants.

Many salt- or boron-affected soils have developed from rocks formed under marine seas or large inland lakes which became saline as they dried up. Other sources of chloride and sodium are irrigation water from bores, wells, dams or rivers, especially at times of low river flow. Although fertilisers do not contribute very much to the development of soil salinity, muriate of potash (potassium chloride), and fertilisers containing it, may add to the problem and are best avoided where salinity is a concern.

Salt injury to container-grown plants is usually caused by high fertiliser rates or use of media or water which is contaminated with salt. In most cases, irrigation is also inadequate. Mushroom composts, sewage sludge, coir dust and ash, all used in the production of soilless media, can sometimes contain high soluble salt levels.

Crops are at greatest risk from salinity in hot weather when water losses through evaporation and plant uptake tend to concentrate salts in the soil solution. Nutrients are also released more rapidly from coated fertilisers (CRFs) as soil temperatures rise and, if there is insufficient leaching to remove the excess, the growing medium will become saline. As a rule of thumb, around 1 to 1.5 times the water-holding capacity of the medium is required in an irrigation to flush salts from soilless media. A single, heavy watering is generally more effective in removing salts than several lighter ones. With drip irrigation, salt will accumulate in the region which is not directly under the emitter and subject to leaching. Capillary rise causes salt to accumulate at the surface of growing media receiving subirrigation. This effect is less when plants are large and can be reduced using mulch mats. Regardless of the irrigation method, an occasional heavy overhead irrigation is beneficial to remove salt gradients within the medium.

Manganese and aluminium can be toxic to crops grown under acid conditions because of their enhanced availability at low pH. Many highly weathered, leached soils are naturally acid, but even very fertile soils become more acidic under the influence of improper cropping, cultivation, irrigation and fertiliser use.

Most nitrogenous fertilisers including ammonium sulphate, urea, ammonium nitrate and mixed fertilisers containing ammonium salts will acidify the growing medium. This occurs because calcium and magnesium are lost when nitrate leaches. Accordingly high fertiliser and leaching rates speed up acidifications. Ammonium sulphate (sulphate of ammonia) is the most acidifying nitrogen fertiliser and should not be used on poorly buffered or low pH soils. Even organic nitrogen sources such as green manure crops, animal manures and sewage sludges will acidify the soil if application rates are greater than the crop requirement.

Nutrient imbalances

As well as requiring sufficient amounts of all the essential elements, plants need them to be supplied in a definite balance between one and another. When an increased supply of one nutrient causes reduced uptake of another, the interaction is said to be antagonistic. An example is the way excess potassium can interfere with the uptake of magnesium or calcium, causing a deficiency of one or both of these elements. In the same way, iron or zinc deficiency can be induced by high phosphorus or manganese. As a general rule, an imbalance of nutrients is most critical when supplies of one nutrient are already marginal for optimum crop performance.

Crop sensitivity and tolerance

Plants differ in sensitivity to specific nutritional disorders. Some plants within the Proteaceae and Fabaceae families, for example, are especially susceptible to phosphorus toxicity and to iron deficiency. Azaleas and rhododendrons are easily burned by nitrogen fertiliser rates needed by other woody shrubs like ilex, cotoneaster, weigela and forsythia for optimum growth. Soft-tissued or rapidly growing crops like orchids, chrysanthemums, carnations, and palms are susceptible to calcium deficiency while *Spathiphyllum*, palms and roses are sensitive to manganese toxicity. Carnations and stocks are prone to boron deficiency, and gerbera, freesia, gladioli and roses to fluoride toxicity.

Plants differ in sensitivity to a specific deficiency for several reasons. Some plants simply need more of a nutrient to maintain physiological processes. Other plants are less efficient at obtaining the element from the soil or of using what they absorb.

Plants can respond to nutrient shortage by growing more slowly (which reduces the requirement for all nutrients), by using absorbed nutrients more efficiently, or by improving root uptake, for example, increased root/shoot ratio, proteoid roots, acidification of rhizosphere, mycorrhizal associations.

Differences in plant tolerance of elements in excess of requirement are related to differences in abilities to exclude elements from uptake at the root surface, to keep absorbed elements away from sensitive tissues, or to render potentially toxic elements less biologically harmful.

CHAPTER 2

IDENTIFYING NUTRITIONAL PROBLEMS

Symptoms expressed by plants in response to a particular nutrient deficiency or toxicity are often sufficiently characteristic to be useful for diagnosis. These visible signs of poor nutrition include stunting, abnormal growth or development of flowers, unusual colours or patterns in the leaves, burns and distortion of individual plant parts.

Strong visual symptoms of deficiency develop when the supply of a nutrient has been severely and abruptly reduced, often at a time of rapid plant growth. Plants grown at a low but stable level of nutrition usually exhibit milder and sometimes quite different deficiency symptoms. However, whenever nutrient supply is inadequate yield loss will usually occur before visual symptoms are readily apparent in the crop.

This has two main implications for diagnosis.

1 Strong deficiency symptoms in a crop are evidence of a recent and abrupt reduction in nutrient supply.
2 The growth or quality of a crop can be significantly limited by nutrient supply even before visual symptoms are seen.

Visual symptoms of nutritional and non-nutritional stress are sometimes confused (Table 1). However, nutritional patterns have certain characteristics which are determined by the way a nutrient is transported within the plant and by its major role in cell metabolism.

Two principal features used when identifying nutritional symptoms are the pattern (and in the case of leaf symptoms, how it relates to leaf venation) and the location of the affected tissues. Nutrients, water and metabolic products are transported in plants through a network of specialised dead and living cells called the 'vascular system'. This system of vessels running through the plant, is most easily seen in the leaves. The physical arrangement of the vascular system and the ease with which individual elements move within the plant (mobility) largely determine where symptoms are found and their appearance. For example, the younger leaves are the first to develop iron deficiency symptoms because iron is not easily remobilised from older tissues in times of shortage. Furthermore, the characteristic yellowing of symptom leaves develops last in tissue near the major veins because this tissue is closest to the supply.

The close relationship between the symptom pattern and the arrangement of veins in the affected leaf is useful for distinguishing nutritional from non-nutritional symptoms.

Table 1 Nutritional disorders and symptoms caused by some non-nutritional stresses which may confuse a diagnosis

Nutritional disorder	Non-nutritional stress
Toxicities	
Sodium, chloride	Heat or moisture stress, air pollution (F, SO_2, O_3).
Nitrate	Frost.
Phosphorus	Herbicide (Roundup).
Ammonium	Root and stem rots.
Foliar nutrient (spray burn)	Some bacterial, fungal and viral diseases.
Deficiencies	
Nitrogen	High or low light, root injury or pot constriction.
Phosphorus	Mild moisture stress, salinity.
Potassium, magnesium	Severe moisture stress or mite damage.
Calcium, boron	Moisture stress (hot dry winds), insect and virus damage of growing points.
Manganese	Root injury.
Iron	Root injury, low soil temperature, herbicide (Roundup).
Molybdenum	Frost.

Characteristics of nutritional symptoms on leaves

- Restricted initially to either young, old or intermediate-aged leaves.
- Patterns are symmetrical and closely related to leaf venation.
- Changes in leaf colour and tissue death develop gradually (rarely overnight).
- Boundaries between green and chlorotic areas on a symptom leaf tend to be diffuse. Strong, definite boundaries are generally produced by herbicides or viruses.
- Leaf patterns are rarely blocky or angular. Such patterns can be caused by a pathogen or occasionally by nematodes.
- Damage to the surface of a symptom leaf is unusual. Nutritional problems impair cell function and rarely cause mechanical disruption of the cuticle.
- Symptoms develop first in tissues most distant from the major veins of the leaf, for example, the interveinal regions, and tips and margins of the leaf blade.

The diagnostic process

When convinced that a problem is nutritional and that pests, diseases and other factors are not involved, the following steps will help you to identify the disorder and, more importantly, to find its underlying cause.

Step 1 – Gathering the facts

Background information is needed to define the problem and to identify the most likely factors contributing to its development. Some causes can be discounted immediately at this stage. Furthermore, details obtained initially, are often useful for establishing the underlying cause of a problem which is crucial when developing an appropriate corrective treatment.

Field description The overall appearance of the crop should be described briefly, emphasising obvious abnormalities such as unusual leaf colouring, stunting or thinning in the crop. Relationships between the distribution of problem areas in a crop and geographical features in a field such as soil type, fence lines and crop rows or, in a nursery, influences such as irrigation bays, walkways and roads, location in the greenhouse including aspect, and proximity to heaters, fans, doorways etc. may point to the cause. Due to the uniform nature of modern potting media, it is unusual for nutritional symptoms to be confined to groups of pots within a production area unless fertilisers were unevenly mixed through the media or topdressed haphazardly. On the other hand, injury caused by herbicides, diseases or insects is characteristically restricted to plants within a limited area.

Severity of problem Estimate the percentage of the crop affected by the disorder and the expected yield reduction. This information may be helpful in deciding if the disorder is the primary problem or a consequence of some other more important stress. Some disorders, such as a mild iron deficiency, may cause only slight yield losses but the symptom could point to more serious problems such as 'root rot', nematodes, waterlogging, and toxicities of various kinds.

Crop Crop name and cultivar are sometimes needed for plant analysis because the interpretative standards differ between crops and may vary between cultivars of the same crop. Also some cultivars are more sensitive than others to a particular deficiency or toxicity.

Crop developmental stage The standards used to interpret plant test results are usually different for the seedling, vegetative or flowering phases and so the testing laboratory must be told the developmental stage of the crop at sampling. Periods of very rapid growth, or changes from vegetative to flowering or seed formation, can have pronounced effects on leaf nutrient composition, as well as the onset of symptoms.

Contributing factors Any change in management immediately before the symptoms developed should be examined as a possible cause. The time taken for symptoms to become fully expressed may also be significant. Symptoms which develop suddenly are generally caused by catastrophic events such as frost, high temperatures, wind, pests, herbicides and chemical sprays. Deficiency symptoms usually develop gradually.

Cropping history Records of previous crops grown on the site, their management and performance are important as residual chemicals in the soil, such as herbicides, liming materials and fertilisers, can affect later crops. Any change in the source or composition of a potting mix prior to the problem should be investigated.

Soil type and depth Relationships between the distribution of affected plants and variations in soil properties within a block should be noted. Affected patches are sometimes associated with land levelling, former roadways, fence lines, irrigation banks, or log burning. Relationships between the distribution of the disorder and soil type or previous land use often indicate soil chemical or physical problems. Examine the soil below plough depth for a possible hard pan, comparing the soil below poor patches of crop with that in better grown areas.

Irrigation type/frequency What method of irrigation is used: overhead, flood, trickle or capillary? A relationship between water distribution patterns and symptoms in a crop may indicate that irrigation is inadequate, excessive or patchy. Severe water stress can trigger early senescence in a crop, the symptoms of which may be confused with nitrogen deficiency or salinity. Overwatering can leach fertilisers away from roots causing deficiency or lead to root death from waterlogging.

Drainage Are the soils well drained? Waterlogging can cause nutritional disorders including manganese toxicity and iron deficiency, or produce wilting, leaf fall, vein clearing and hormone-like symptoms on new growth.

Weather conditions Were weather conditions unusual around the time the problem developed? Periods of heavy rain, drought, high or low temperatures, high humidity or frost may all cause a setback in the growth of a crop. Both current weather and that occurring some weeks prior to the appearance of symptoms may be significant.

Fertiliser history Problems can sometimes be traced to a recent change in fertiliser practice, to a history of over or under use, or to an unbalanced fertiliser program. Current fertiliser programs should be compared with what was previously used and with the practices of other growers in the area.

Block history Does the problem area coincide with a previously cropped or fertilised section of the block? Compounds excreted into the soil by some crops including chrysanthemum, stocks, sunflower, walnut, erigeron, tree-of-heaven, *Juniperus*, *Quercus*, *Pinus* can inhibit the growth of a following crop. Cropping the soil with the same species over an extended period can also lead to problems. Residual fertiliser in soil may be detrimental to following crops, for example, when phosphorus-sensitive plants are grown on orchard or vegetable land with a long history of high phosphate application.

Spray program What chemical sprays have been used on this crop – pesticides, nutrients or other? Pesticide and nutrient sprays can burn foliage or flowers if not used appropriately. Nutrient sprays and some fungicides containing trace elements can contaminate samples collected for plant testing and confuse a diagnosis. Recent use of herbicides on neighbouring land should also be noted as spray drift can injure crops hundreds of metres from the point of application.

Plant health Are plants diseased or infested with insect pests which damage the roots or the vascular system? Pests or disease can produce nutritional symptoms and alter the nutrient composition of tissues sampled for plant analysis. Insects and many common pathogens also cause injuries that are confused with nutritional symptoms. Often the causal organism or debris from its activity (eggs, shells, or fungal hyphae or fruiting bodies) can be seen using a simple hand lens.

Step 2 – Diagnosis from visible symptoms

All visible evidence of a deficiency begins as a deterioration of processes in a plant's cells. Boron deficiency leads to death of growing points and distortion of leaves and flowers because boron is needed for the proper regulation of cell division. Similarly, nitrogen and magnesium deficiency cause chlorosis or yellowing of leaves because these nutrients are essential for synthesis of chlorophyll, the green pigment in leaves.

Links between the biochemical role of a nutrient and the symptom which results when it is deficient, are common in plants. This is why it is possible to make generalisations about symptoms and why visual diagnosis can be useful, even with unfamiliar crops.

The two most important diagnostic features of a nutritional symptom are where the symptom is found on the plant (location) and its appearance (colour and pattern).

Description of symptoms Having previously described the general background of the problem (field description), it is now necessary to look closely at the symptoms on individual plants. This is best done in the subdued light of early morning, late afternoon, or an overcast day. In full sun, light scattered from the leaf surface tends to obscure some of the more subtle effects.

Location Where do the symptoms first appear on the plant? Nutritional symptoms rarely develop uniformly over a plant but show first in a specific organ such as the leaves, flowers, roots or shoot. Additionally, leaf symptoms are usually confined to the upper, middle or lower region of a shoot, with the location determined primarily by the mobility of the deficient element.

Mobile elements like nitrogen, magnesium or potassium are moved about the plant relatively easily so, when there is a shortage, reserves in older tissues can be used to satisfy actively growing tissues such as new shoots or developing flowers. This is why the older leaves develop symptoms before younger leaves when one of these mobile elements becomes deficient.

Other elements, such as iron, boron or calcium, move less readily from old to younger tissues. They tend to accumulate in older tissues and, when deficient, give rise to symptoms initially in the newer or upper leaves or in developing flowers. Because the plant cannot draw on internal reserves of these immobile elements, even a temporary shortage can severely restrict or distort the growth of young tissues.

Symptoms of nutrient toxicity generally show first in the oldest leaves because they are the major sites of nutrient accumulation. Young tissues have less developed vascular connections and lower rates of transpiration and so acquire nutrients from the transpiration stream less effectively.

Pattern The size and form of the plant, its overall foliage colour as well as the colour of symptom leaves, the pattern of chlorotic (pale) or necrotic (burnt) areas in relation to vein pattern, or the irregular shape of affected organs may all help to identify a disorder.

Deficiency or toxicity? The first question to answer is: Do the symptoms indicate a deficiency or a toxicity? The following generalisations concerning deficiency and toxicity symptoms may help answer this question.

- Deficiency symptoms are typically restricted to a particular age of tissue (young, recently mature or old) unless more than one problem exists at a time.
- Toxicities commonly produce necrotic (tissue death) symptoms. Necrosis on new leaves almost always indicates a toxicity (an exception is leaf tip burn caused by calcium or boron deficiency). Burns or necrotic spots on old leaves may or may not be due to toxicity – some deficiencies, notably those resulting from a severe shortage of potassium or magnesium, can also produce necrosis on older leaves.
- Symptoms on both old and new leaves usually indicate a toxicity. When an excess of one element causes a nutrient imbalance, deficiency symptoms may be seen in the young leaves while older leaves may show a burn or other toxic symptoms. For example, toxicities of phosphorus, manganese or zinc can induce iron chlorosis in the young leaves as well as symptoms of nutrient excess in the old leaves.
- Toxicity symptoms usually appear suddenly and may rapidly worsen with time. Strong symptoms are often apparent as early as a day after the plant has been stressed.

Which deficiency? If a deficiency is suspected, the location of symptoms on the plant is a useful guide to whether the responsible element is mobile or immobile. The way symptoms develop on leaves, shoots or flowers provides further clues. Small, irregular-shaped leaves, shortened internodes, aborted flowers and poor seed set can be characteristic of a particular deficiency.

Diagnostic keys like the one in Table 2 provide a framework for a visual diagnosis.

Although visual symptoms are very helpful for diagnosis, the approach does have three major weaknesses.

- Clear visual symptoms usually only appear when a disorder is quite advanced and some loss of yield or quality has occurred in a crop. By this time, even prompt remedial action will not restore the loss.
- The absence of symptoms does not mean that crop nutrition is adequate as yield or quality can be reduced before symptoms have been expressed. This condition is called ‘hidden hunger’.
- Visual symptoms can be unreliable for diagnosis when more than one element is limiting.

Table 2 Quick guide to nutrient deficiencies – what to look for

Symptoms first seen in *older* leaves	
Leaf coloration even over whole leaf	
Nitrogen	Pale green to yellow leaves.
Phosphorus	Leaves dull, lacking lustre, bluish green or purple colours. Poor growth.
Leaf coloration forms a definite pattern	
Potassium	Scorching and yellowing, commonly around the edges of leaves, which may become cupped.
Magnesium	Patchy yellowing often with a triangle of green remaining at the leaf base. Less commonly, grey or light brown scorching with severe deficiency in some species.
Symptoms first seen in *young* leaves	
Leaf coloration forms a pattern	
Iron	Almost total loss of green between veins, leaving faint green 'skeleton' of veins on leaf.
Zinc	Severe restriction of leaf size or stem length, or both (hence the terms 'little leaf' or 'rosetting'). Distinct creamy yellow interveinal pattern. Distorted young leaves.
Copper	Tips of leaves cupped, narrow, distorted or scorched. Defoliation from tip. Chlorosis interveinal or irregular.
Symptoms first seen in either *old* or *young* leaves	
Leaf coloration forms a pattern	
Manganese	Mottled diffuse pale green to yellow patches between veins. No restriction of leaf size (unlike zinc).
Symptoms usually most prominent in other tissues – seen first in youngest tissues and fruit	
Calcium	Death of growing point or margin of young leaves.
Boron	Internal cracking or breakdown of stem or roots. Surface cracking, corkiness of midrib or petiole. Death of growing point and multiple buds. Aborted flowers.

Step 3 – Confirming the diagnosis

Where there is uncertainty about the visual diagnosis or, where an incorrect diagnosis would prove costly, other techniques can be used to confirm the diagnosis.

Trial treatment of a portion of the crop This is the most direct method of confirming a diagnosis and can be a good way of checking that fertiliser usage is adequate. Providing some part of the crop is left untreated to gauge treatment effectiveness, this method gives the most useful answer. But trials can be slow and, if unsuccessful, provide no new information which could lead on to a correct (successful) diagnosis.

Plant tissue analysis Tissue analysis is a powerful diagnostic tool which provides good direct evidence of the nutritional status of a crop. It can be used to verify a visual diagnosis and, because it gives information about the availability of a wide range of nutrients, it enables a new interpretation if the original symptom-based diagnosis is wrong. However, plant analysis alone may not reveal the underlying cause of a nutrient disorder. Quick sap tests of leaves or other tissues are usually specific for one element and can be unreliable if they have not been calibrated for the crop.

Plant analysis is used to

- Diagnose nutrient deficiencies and toxicities:
  - confirm a diagnosis based on symptoms,
  - identify 'hidden hunger', and
  - suggest additional tests to identify a problem.
- Predict nutrient disorders in current or future crops.
- Develop and adjust fertiliser programs.
- Measure the amounts of nutrients removed in crop produce or residues with the view to replacement.
- Survey nutrient status of a crop throughout a district.
- Compare the nutritional status of soils or growing media.
- Estimate the dietary value of a crop.

Soil and water analysis These are useful aids in determining the cause of crop nutrient problems, but by themselves are of limited value for identifying disorders. Although soil analysis is the only practical means of forecasting crop nutrient needs prior to planting, the test can be misleading if it is not calibrated for the crop and for the soil or growing medium used.

As no one diagnostic procedure is entirely satisfactory, it is good practice to support a diagnosis with both laboratory tests and field observations.

Step 4 – Correcting the problem

Corrective treatments should be directed at solving the underlying causes of a problem and not simply at reducing the immediate effects on the crop. For example, liming the soil would not be the best way of correcting a calcium deficiency if this had been induced by excessive use of potassium fertilisers. In this instance, a series of calcium sprays would be a useful way of minimising damage to the current crop but long-term relief could only be obtained by reducing potassium usage.

Step 5 – Following up

The only sure way of knowing whether a correct diagnosis has been made and of improving diagnostic skills is to see whether the affected crop has responded to treatment. Unless the real reason for a crop setback is discovered, the basic problem may continue to worsen or reappear at some later time when the grower will be no better equipped to deal with it.

CHAPTER 3

PLANT ANALYSIS

Plant analysis is useful for assessing the nutrient status of a crop. This is done by comparing nutrient concentrations in a specific plant tissue (usually leaves) with predetermined ranges (leaf standards) for healthy, productive crops of the same species. Although plant analysis allows nutritional problems to be identified before they become severe (prognosis), the procedure is more often used to confirm the identity of a fully developed disorder (diagnosis).

Plant analysis tells us how well a crop is presently supplied with nutrients but not whether supplies will be adequate in the future. It, therefore, differs from soil analysis which is used to predict the soil's ability to supply nutrients over the life of a crop.

The concentration of a nutrient in plant tissue when that nutrient just begins to limit growth (or quality) is called the critical level. This term is usually defined 'as the tissue concentration associated with a 10% reduction in yield due to deficiency or toxicity' (Figure 2). Nutrient ranges in tissue when nutrient supply is deficient, low, normal, high or toxic have also been established for many crops (see box on page 17). These ranges describe nutritional status from deficiency through normality to toxicity better than is achieved with a single critical value. Whereas critical levels are concerned only with yield or quality loss, deficiency or toxicity, ranges indicate the likelihood of symptoms being seen in the crop. For ornamental crops, this is particularly important because their commercial value is tied closely to their aesthetic appeal.

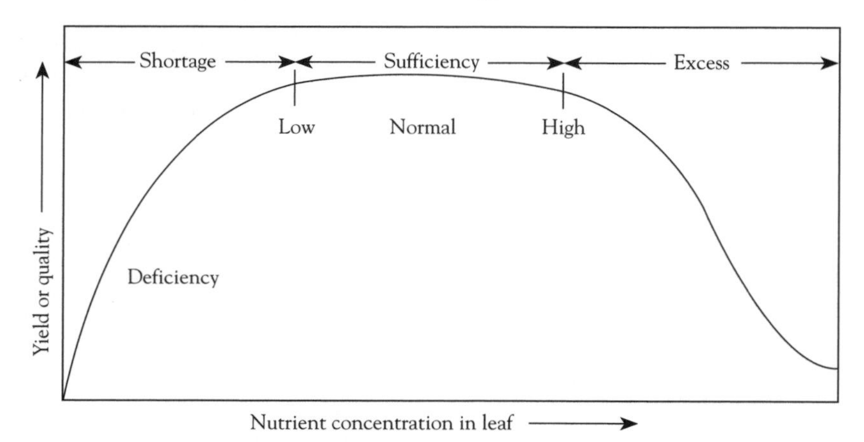

Figure 2 The relationship between plant yield or quality and leaf nutrient concentrations

The five classes of plant nutrient status normally defined are:

- **Deficient** Symptoms present – nutrient is too low for optimum performance.
- **Low** No symptoms but nutrient is too low for optimum crop performance.
- **Normal** No symptoms – level is adequate.
- **High** No symptoms – level higher than necessary which may cause imbalance or loss of quality.
- **Excess** Level too high for optimum performance – toxicity symptoms may be present.

Leaf analysis standards used by this laboratory (NSW Agriculture) are provided as Appendix 1. Guidelines for some of the crops have been developed from a subjective assessment of diagnostic and research analyses. In such cases the 'Normal' range refers to analyses for healthy productive crops with no symptoms. The 'Deficient' range refers to analyses for crops with deficiency symptoms (or otherwise established to have been deficient), including analyses from research trials. This range is probably not completely defined by the data listed. The 'Excess or Toxic' range includes data from crops either showing toxic symptoms or where a serious nutritional imbalance was suspected.

'Below Normal' and 'Above Normal' replace the 'Low' and 'High' ranges which were used in the tables for which more complete information was available. Although these values are not considered 'Normal', loss of yield or quality due to nutrition has not been definitely established.

Sampling for plant tissue analysis

The nutrient composition of a leaf changes during the season and varies with its age and position on the shoot, as well as the developmental phase of the plant (vegetative, flowering, fruiting, tuber development). Taking the correct type of leaf at the right stage of the crop's growth is, therefore, critical in using plant analysis. **Leaves must be sampled correctly or the interpretative standards will not apply and the analytical results will be misleading.** If this is not possible, a detailed description of the sampling procedure used will assist an experienced diagnostician make an interpretation.

Sampling method General directions for most ornamental crops. Detailed procedures for individual crops are given with the plant analysis standards in Appendix 1.

Tissue Whole leaves, that is, blade (lamina) including midrib, but often removing extended petioles. Where the leaves are compound, discard the main petiole, retaining the short stalks attached to the individual leaflets. Sample only healthy plants of normal growth characteristics.

Leaf age A fully expanded, recently matured leaf.

Number of plants Take leaves from about 25 plants of the same age, throughout a uniform and representative section of a block within one soil type, or from containers receiving a similar nutrient program, following a zig-zag pattern throughout the sampled area.

Number of leaves per sample 60 to 100

Making a two-sample comparison Where no leaf standards exist or their use is inappropriate because sampling could not take place at the correct time or in the specified manner, some useful guidance in diagnosing a nutritional disorder can also be gained by comparing the nutrient levels in leaves from healthy and affected plants. When this is done, both the plants and the leaves in both samples must be at the same developmental stage.

Quick sap tests Sap testing is a more rapid and less expensive means of identifying some nutritional disorders than leaf analysis. Speed is important with ornamental crops because they mature quickly and rapidly lose value if a treatment is delayed.

With sap testing, the concentration of a nutrient in fluid expressed from leaves or petioles provides a guide to how well the crop is supplied with that nutrient. These chemical tests can be done either in the laboratory or, more roughly, in the field by the grower using a selective ion electrode or relying on a colour reaction in a tube of solution or on a treated paper strip.

One popular test method (Merckoquant) uses paper strips impregnated with chemicals sensitive to nitrate. The paper is wetted with fluid extracted from plant tissue and the nutrient concentration is estimated from the intensity of the colour which develops within a given time. The test is interpreted by comparing the recorded value with standard nutrient ranges for healthy plants.

The concentrations of soluble nutrients in sap are far less stable than the total nutrient concentrations in tissue used for plant analyses. Sap testing tends to be highly sensitive to the current status of nutrient supply; Leaf analysis tells us more about the nutritional history. Sap concentrations can rise and fall very quickly after a fertiliser application, so the assessment of crop nutrient status is strongly influenced by when the test was done. The result can also be influenced by non-nutritional factors including the time of day when the sample was taken, the degree of shading of the sampled leaf, the amount of cloud cover, water stress and the time since the last irrigation. Levels of one nutrient in the sap can be influenced by a deficiency of another. For example, high nitrate is used as a marker of molybdenum deficiency. As quick tests provide information about a single element only, interactions like this may go unnoticed. Furthermore, if the diagnosis is wrong, the test provides no additional information upon which to make an alternative diagnosis. Apart from the difficulty of controlling the testing conditions in the field, large errors may come from the selection of a representative tissue sample, particularly where plant growth is highly variable. Finally, sap testing cannot be used with many woody plants because it is too difficult to obtain sufficient sap from leaves or petioles.

Sap testing is most reliable when it is used to monitor crop nutrient status in conjunction with plant analysis. With regular use on particular crops, growers can become familiar with the tests' shortcomings and avoid major errors in interpretation. Sap analysis and other quick tests for nutrients in soils and drainage water should become more important in ornamental horticulture as these industries strive to improve efficiencies in fertiliser use and minimise nutrient losses in waste water.

CHAPTER 4

COMMON NUTRITIONAL PROBLEMS AND THEIR CORRECTION

Deficiencies of the major elements, nitrogen, phosphorus and potassium, are encountered in most crops whenever insufficient fertiliser is used. Phosphorus deficiency is more likely on uncropped soils, and nitrogen deficiency where there has been excessive leaching or strong immobilisation (nitrogen drawdown). Potassium deficiency is usually encountered on heavily cropped, light soils when potassium is not regularly replenished by the fertiliser program. Ornamental crops are subject to a wide variety of disorders because less is known about their needs and because soilless growing techniques are often employed. Deficiencies of iron, magnesium, boron, calcium, manganese and sulphur, and toxicities of nitrate, ammonium, boron, manganese and phosphorus are not unusual.

Some nutrient disorders are caused by an inappropriate pH. Soils which have been cropped for some time can become quite acidic, leading to toxicities of manganese and aluminium or deficiencies of magnesium, calcium and molybdenum. Deficiencies are most often the result of an inadequate supply but can also be caused by a lack of balance between available nutrients, or even over-fertilisation, particularly the excessive use of nitrogen, phosphorus or potassium. Iron deficiency is found on alkaline or over-limed soils or media, where excess phosphate has been used, and in crops grown in soilless media which are low in iron.

Some idea of the types and relative importance of specific nutritional problems encountered in ornamental crops can be obtained from diagnostic leaf analysis records. In Table 3 the relative frequency of specific nutritional disorders which were identified in crops from 1982 to 1990 is shown.

The records show that of the 99 carnation samples submitted to the laboratory for diagnosis in that period 59% were subsequently found to be suffering some nutritional imbalance. High leaf zinc was the most common deviation from the normal range – 56% of all cases. Nitrogen deficiency was confirmed in 49% of cases and high leaf manganese in 32% of cases. The high concentration of zinc, manganese and copper in carnation leaves is probably more a reflection of the wide use of fungicidal sprays containing these elements than of toxicity due to root uptake. On the other hand, the importance of nitrogen deficiency could be underestimated at 49%. This disorder is readily identifiable from symptoms and so fewer samples are submitted for diagnosis.

Table 3 Relative importance of common nutritional abnormalities identified by leaf analysis in selected ornamental crops. Table based on diagnostic records from the Biological and Chemical Research Institute (NSW Agriculture) from 1982 until 1990

Crop	Percentage occurrence[1] (%)																		Number of samples	Percentage affected (%)
	below normal									above normal										
	N	P	K	Ca	Mg	Mn	Cu	Zn	Fe	Cl	N	P	K	Mg	Na	Mn[2]	Cu[2]	Zn[2]		
Carnation	49				11				14			19				32	18	56	99	59
Chrysanthemum	36					17			22		22					22		78	40	90
Rose	49	12	26	12	23	10	24		15		57	11	17	26				39	85	88
Palm	23	13	60	54	37	25	17	21	23							27			62	84
Protea	60	63		33			73	17					22	25	33	28		13	61	98

[1] The number of samples exhibiting leaf nutrient levels above or below the normal range for the crop expressed as a percentage of the total number of samples.

[2] Many samples showing high levels of manganese, copper or zinc were affected by residues of pesticides on the leaf surfaces.

The large number of palm disorders identified in this period shows that this crop's nutrition was poorly understood by growers. However, it is also possible that the leaf analysis standards used for diagnosis were not entirely appropriate. Were the optimum levels for some nutrients set too high for this crop? This was almost certainly the case for the various *Protea* species which were examined and for which only very limited and tentative standards are available even now.

Developing a fertiliser program

The great diversity in plant species and conditions for growing ornamental plants makes it difficult to generalise about optimum fertiliser rates. This information is simply not available for most crops and, where it is, tends to be specific to the prevailing cultural conditions. On-farm research is required to achieve maximum efficiency in fertiliser use.

In this section, we will explain how to establish the fertiliser needs of a new crop and how to check the adequacy of an existing fertiliser program.

Finding the optimum rate for a new crop The experimental design described here, allows for four fertiliser rates; more may be used but this would increase the size of the trial considerably. Choose rates which are lower and higher than what you think the ideal rate will be, keeping in mind that the nutritional conditions giving the fastest growth do not always give the best product. Quality concerns such as hardiness, foliage colour and flowering are often more important than vigorous growth. In the example, we have chosen 2, 4, 6, 8 kg/m^3 of a controlled release fertiliser with 16%N, as experience has shown that this range covers the needs of most container plants.

Each treatment should be carefully applied to 5 pots in which plants of uniform age, size and type will be planted. The total number of pots in the trial will be 4 treatments × 5 replicate pots = 20 pots.

These treated pots should be set out in the production area as indicated (Figure 3) so that shading, uneven watering and other external influences on growth are experienced equally by all treatments. Plants are then grown until there are clear differences between treatments. Ideally, the trial should be run until plants reach a saleable size. At harvest, the growth and quality of each plant is assessed. Growth can be estimated by weighing the plant tops, however, other non-destructive measurements, such as height, girth and flower numbers, may be more appropriate depending on the product.

The performance of plants at each fertiliser rate is then estimated by averaging the results for the 5 replicate pots. When these values are plotted against the appropriate fertiliser rate, a curve similar to that shown in Figure 4 will be obtained. From this data, it is relatively easy to pick the optimum fertiliser rate for the crop. It is also clear from this response that rates below the optimum will give inferior growth, and rates above it may reduce or delay flowering, reduce quality or simply add to the cost of production and increase nutrient losses into the environment.

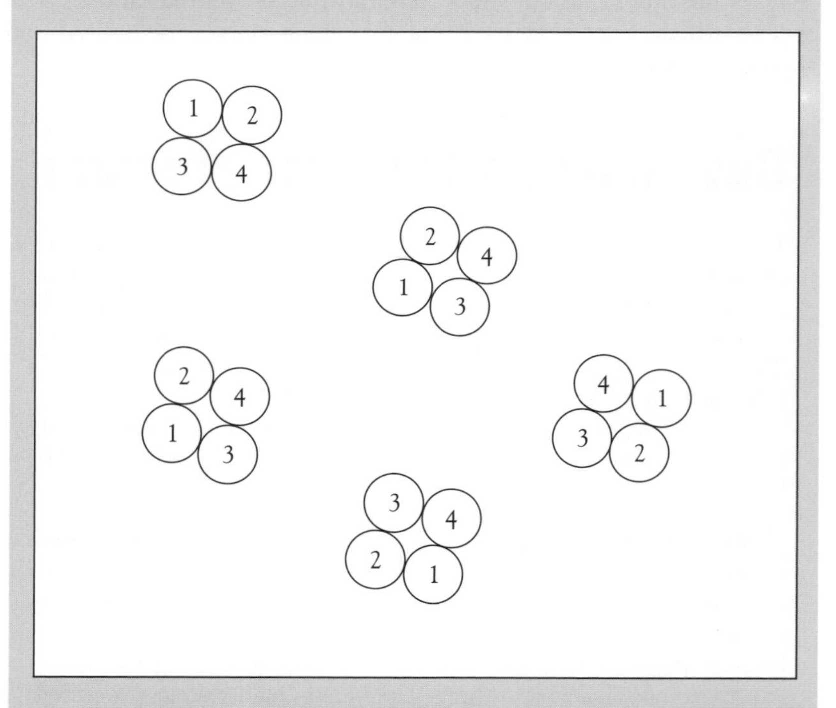

Figure 3 Recommended arrangement of pots in a trial with four fertiliser treatments and five replicate pots

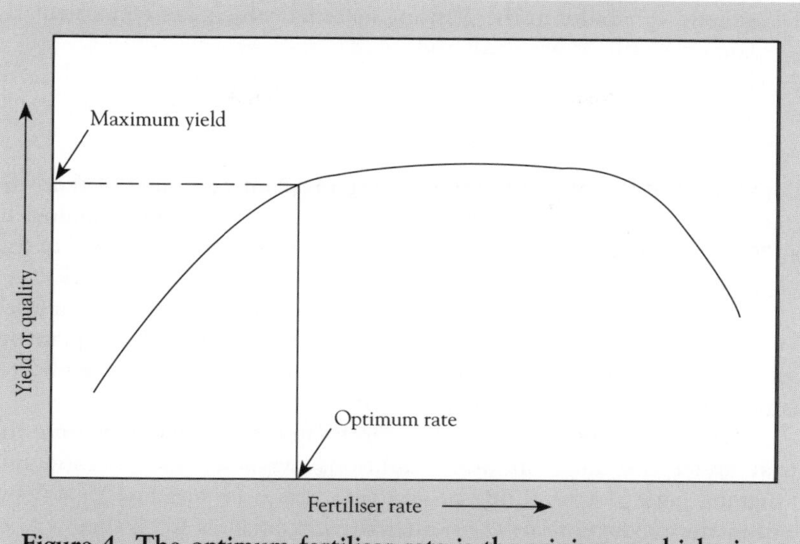

Figure 4 The optimum fertiliser rate is the minimum which gives the desired yield and quality

Checking an existing fertiliser program

Plant analysis can be used to check the efficiency of a fertiliser program. This test will show whether the crop is receiving too little or too much fertiliser and whether the fertiliser has an appropriate nutrient balance. However, a plant analysis will not identify the ideal fertiliser rate. This can be found by observing the growth of plants in a fertiliser trial.

The 20% test One way of simultaneously evaluating and fine-tuning a nutritional program is to compare the performance of plants receiving 20% more and 20% less than the current fertiliser rate. As in the previous experiment, the three treatments – current rate less 20%, current rate and current rate plus 20% should be replicated by 5 pots and these should be arranged in the production area to minimise external influences on growth.

Any marked improvement in growth or quality achieved from varying the current fertiliser rate indicates that the fertiliser program can be improved. This type of trial should be repeated in summer and in winter to show the influence of climate on plant fertiliser needs.

Trial design

In any fertiliser trial, the growth and quality of plants can be significantly influenced by factors other than the fertiliser treatments. The two main influences are:

1 genetic variability in the planting material, which can be particularly large if seedlings are used; and
2 lack of uniformity in the growing conditions. Uneven shading or watering is very common in nursery production areas especially if overhead sprinklers are used.

Both factors can have a greater influence on plant performance than the treatments under evaluation, and this can give rise to a wrong finding. In a properly designed trial, strategies are used to reduce the likelihood of this happening.

The influence of genetic variability in growth rate can be minimised by replicating (repeating) treatments and by using uniform planting material (same cultivar size and maturity) in the trial. Five to 10 replicates is normally adequate for a fertiliser trial.

The second distortion, climate, can be minimised by conducting the trial under the most uniform conditions available and by randomly assigning pots (not favouring or grouping pots from any one treatment) within the production area. An alternative strategy is to arrange pots in groups containing one plant from each of the treatments. This will ensure that any shading or watering differences that affect growth will be shared equally by all treatments.

NITROGEN

Plants have a higher requirement for nitrogen than for any other nutrient (with the exception of potassium in some crops). Fertilisers or manures containing nitrogen are needed to maintain the fertility of nearly all soils which are being cropped. Fertilisers are even more important with soilless media which generally have little inherent nitrogen fertility and often a significant biological requirement for nitrogen ('Nitrogen drawdown'). Digging sawdust, straw and other woody crop residues high in carbon into the soil may temporarily immobilise nitrogen and starve the crop. Frequent and often excessive irrigation of container plants in nurseries relying on overhead sprinklers also leads to substantial leaching losses of nitrogen (15 g N/m^3/month). Nitrate leaching caused by irrigation or rainfall can also be severe in soils especially those which are light, sandy and low in organic matter. Gaseous losses of nitrogen (volatilisation) can be large when fertilisers are surface applied or incorporated into a growing medium which is cold, waterlogged or has been recently limed.

In containerised nursery stock, nitrogen deficiency is usually caused by inadequate fertiliser rates, excessive leaching of soluble fertiliser, use of media with high N drawdown, too rapid a release from controlled release fertilisers, or the failure to supplement the nutrition of liquid fed plants on removal to a retail outlet.

When all other nutrients are adequate, the supply of nitrogen fertilisers can be used to control plant growth. Restricting nitrogen supply is a common method for conditioning (hardening) seedlings for planting out, acclimatising plants for indoors, and for delaying the development of containerised plants for a market. Plants which have been stressed in this way appear pale and stunted but quickly regreen and resume normal growth when resupplied with nitrogen.

Plants can have quite different nitrogen needs and this is generally a reflection of how fast they grow; the fastest growing plants have the highest requirements. The need for nitrogen is greatest in all plants when they are growing rapidly and reduces when growth is slowed by unfavourable conditions such as low light, cool weather or water stress. Leafy, fast-growing crops under high light generally have the greatest need for nitrogen.

Nitrogen nutrition can influence the quality as well as the quantity of production. Such things as plant form, height and colour, flower colour, size, number and post harvest life, and pest and disease resistance, as well as hardiness to cold and drought are all determined to some degree by the availability of nitrogen.

Function Nitrogen is an essential constituent of protein and chlorophyll, the green pigment in leaves. It is quite mobile within the plant and, when supplies are limiting, the needs of new growth can be satisfied from older tissues. As a consequence, strong leaf symptoms of deficiency show first in older leaves.

Symptoms Nitrogen-deficient plants are generally spindly and stunted with pale green foliage. Branching is reduced due to the inhibition of lateral shoots but root growth is often stimulated. The oldest leaves are first to lose their green colour and this uniform yellow chlorosis spreads up the shoot until all leaves are pale green or yellow. In some crops, including roses, the veins of older leaves may clear before the interveinal tissue. However, in all instances, the older leaves eventually turn bright yellow and fall prematurely or wither on the plant. Red or purplish colours can appear on the chlorotic leaves and stems of some plants including roses, azaleas, gerberas, waratah, cissus, marigold and celosia. Flowers produced by deficient plants normally are small and few in number on short, thin stems.

Too much nitrogen can delay plant maturity, promote excessive vegetative growth, weaken flower stems and reduce the post harvest life of both flowers and potted plants. Plants receiving too much nitrogen often have poorly developed roots in relation to the tops. These plants can also be soft and sappy which makes them less hardy and more susceptible to pests and diseases. Heavy nitrogen use can increase the calcium requirement of a crop and induce deficiency by stimulating excessive growth. Deficiencies of calcium are more likely when ammonium fertilisers are used at high rates because ammonium also interferes with calcium uptake by roots. Ammonia released by fertiliser or manures especially when applied to cold wet soils, can burn or kill seedlings. The resulting injury to roots and lower stems enables fungal and bacterial rots to establish. Excess ammonium causes an interveinal chlorosis of lower leaves in roses. Symptoms of ammonium toxicity in chrysanthemum include reduced top and root growth, necrotic spots on middle to lower leaves, thickened leathery leaves, browning of roots and, ultimately, plant death. Thin, leggy growth of plants growing under low light may be prevented by reducing nitrogen supply. Fertiliser rates can be reduced by as much as 50% of the outdoor rate due to the reduction in growth which occurs when plants are brought indoors.

Corrective treatments Total application rates of nitrogen fertiliser for field-grown crops vary greatly (Table 4) from 25 kg N/ha to 300 kg N/ha depending on the crop, soil, rainfall, spacing, cropping level, fertiliser form and previous soil management including the use of manures. Recommendations for containerised plants are also extremely variable (Table 4) being influenced by plant species, fertiliser form (organic, coated or soluble) and application frequency, irrigation method and degree of leaching, growing medium, and environmental conditions. A procedure for assessing the adequacy of a current fertiliser program or for developing a new program is given (see page 21).

Micro-organisms can consume (immobilise) as much as 50% of the fertiliser nitrogen in potting mixes which have a high carbon to nitrogen ratio (>10:1). Unless additional fertiliser is added this competition for

Table 4 Range in published fertiliser recommendations for ornamental crops[1]

Crop	N	P[2]	K[2]
Field grown[3]		kg/ha	
Cut flowers	25–300	25–200	60–250
Bedding plants	60–100	22–70	68–220
Container grown[4]		g/m^3/month	
Potted colour	130–300	20–80	65–160
Foliage	100–400	30–130	60–250
Woody shrubs	75–270	10-90	40–180
Phosphorus-sensitive plants	6–55	2–15	5–25
Propagation material	55–170	10–30	35–100

[1] Summary of fertiliser recommendations from several sources including published and unpublished research results and grower experience. The appropriate rate of fertiliser is determined by many factors including crop, production period, fertiliser form, growing medium, irrigation method, light and temperature conditions.
[2] Rates are expressed as the elemental form not as P_2O_5 and K_2O.
[3] This range covers rates for soluble and controlled release fertilisers (CRFs). Recommendations for CRFs are generally 50% lower than for soluble fertilisers.
[4] Recommended rates for controlled release fertilisers.

nitrogen can cause a deficiency in plants. Wood waste-based media need around 1.5–2.9 mg/L N per week for each 0.1 NDI unit drop below 1. NDI is an index of nitrogen fertiliser demand based on a simple incubation test (Australian Potting Mix Standard AS 3643–1993). Nitrogen fertilisers commonly used in ornamental plant production are listed in Tables 5 and 6.

Flowering crops generally require most of their nitrogen in the latter half of the growing period. It is, therefore, normal practice to supply only part of the nitrogen requirement at sowing or planting time and the remainder which may be up to two-thirds of the total as two or more side dressings through the season. Splitting the fertiliser application in this way avoids burning and keeps leaching and volatilisation losses to a minimum. Nitrogen can also be supplied after planting through an irrigation line (fertigation). Liquid fertilisers are best applied at low rates, but they should be applied frequently to ensure an even level of nutrition. Nitrogen forms which are readily water soluble, such as ammonium nitrate, potassium nitrate and urea, are mainly used for this purpose. Nitrate measurements on sap and leachate from a potting mix can be used as a guide to when crops need liquid feeding.

Nitrogen deficiencies can be quickly rectified with foliar sprays but this cannot be the primary means of supplying nitrogen to the crop because leaf uptake is limited. Urea is the preferred nitrogen form for foliar application because it is safe to use at the high rates needed to supply nitrogen through leaves. Low biuret urea (<0.4%) should be used to avoid phytotoxicity.

Controlled release fertilisers (CRFs) and liquid fertilisers have largely replaced traditional straight and compound inorganic fertilisers as the major nitrogen source in container plant production. CRFs are favoured because they release nutrients at a slower rate than conventional soluble fertilisers, giving a more even supply of nitrogen to the crop during the

Table 5 Inorganic fertilisers and chemicals used in the production of ornamental crops

Fertiliser	Nutrient content	kg of fertiliser needed to supply 1 kg of N, P or K	Advantages	Problems
Urea	46% N	2.17 kg	High analysis – low freight, low cost.	Losses if left on surface. Water or cultivate in.
Ammonium chloride	26% N	3.85 kg		Chloride content may be a problem in saline conditions.
Ammonium nitrate	34% N	2.94 kg	Low cost. Quick response. Half N as nitrate which moves quickly to the roots.	Nitrate can leach easily.
Ammonium sulphate	21% N	4.76 kg	Not recommended except for alkaline soils.	Expensive. Highly acidifying.
Calcium nitrate	11.9% N	8.40 kg	Highly soluble. Source of Ca.	Expensive. Nitrate can leach.
Diammonium phosphate (DAP)	18% N 20% P	5.55 kg for N 5.00 kg for P	Supplies both N and P.	Strongly acidifying. Concentrated source of N and P but lacks Ca.
Mono ammonium phosphate	11.8% N 26% P	8.47 kg for N 3.85 kg for P	Source of N and P.	
Dipotassium phosphate	44.9% K 17.8% P	2.23 kg for K 5.62 kg for P	Supplies both K and P.	Expensive.
Magnesium nitrate	10.9% N	9.17 kg	Soluble source of Mg and N.	Expensive. Nitrate can leach.
Potassium sulphate	41% K	2.44 kg	Free of chloride.	Higher cost compared to muriate of potash.
Potassium dihydrogen phosphate	28.7% K 23.5% P	3.48 kg for K 4.26 kg for P	Source of K and P.	Expensive.
Potassium chloride (muriate of potash)	50% K	2.00 kg	Cheapest form of K. Highly soluble.	Has a high chloride content. Do not use where soil salinity is a problem.
Superphosphate: single	8.8% P	11.36 kg	Supplies P, Ca and S.	Relatively low analysis.
double triple	17% P 36.5% P	5.88 kg 2.74 kg	Concentrated sources of P giving savings in freight and handling.	

Table 5 Inorganic fertilisers and chemicals used in the production of ornamental crops (continued)

Fertiliser	Nutrient content	kg of fertiliser needed to supply 1 kg of N, P or K	Advantages	Problems
Potassium nitrate	38% K 13% N	2.63 kg for K 7.69 kg for N	Supplies both N and K in soluble form.	Expensive. Nitrate can leach.
Sodium nitrate	16.5% N	6.06 kg		High Na content.
N–P–K mixtures:				
solid	Various proportions of N, P and K	–	A convenient way to supply N, P and K in one application.	More expensive than single element fertilisers.
controlled release	Various proportions of N, P and K		Longer lasting. Lower leaching losses.	More expensive than normal fertilisers.

Table 6 Organic fertilisers used in the production of ornamental crops

Fertiliser	Nutrient content	kg of fertiliser needed to supply 1 kg of N, P or K	Advantages	Problems
Blood and bone[1]	5% N 4% P	20 kg for N 25 kg for P	Supplies both N and P in long–lasting forms.	Large quantities needed. Expensive.
Fish meal	10% N 2.5% P	10 kg for N 40 kg for P	Supplies both N and P in long-lasting forms.	Large quantities needed. Expensive.
Hoof and horn	13% N	7.7 kg for N	Rich source of N.	Expensive.
Poultry manure (high grade)	3.3% N 2.0% P 1.5% K	30 kg for N 50 kg for P 67 kg for K	Nutrients in longer lasting forms.	Large quantities needed. Readily leached.
Poultry manure (fair grade)	2.0% N 1.2% P 1.0% K	50 kg for N 83 kg for P 100 kg for K	Nutrients in longer lasting forms.	Large quantities needed. Readily leached. Nutrient content variable.
Sewage sludge	2–4% N 1.5–6% P	25–50 kg for N 16.7–66.7 kg for P	Nutrients in long-lasting forms.	Expensive. Can be contaminated with heavy metals and pathogens.
Compost	1.5–3% N 0.3–2% P	33.3–66.7 kg for N 50–333.3 kg for P	Nutrients in longer lasting forms.	Expensive. N drawdown can be high.

[1] Typical analysis: the analysis of blood and bone varies from brand to brand.

growth period and reducing the risk of burning and leaching losses. Similar benefits can be obtained with liquid feeding programs which allow fertilisers to be metered out in regular, small amounts over the life of a crop. With both these methods, the supply of nitrogen can be better matched to crop demand, reducing the risk of burn, fertiliser wastage and the environmental hazard from nutrient runoff. Most liquid feeds contain from 100–300 ppm of nitrogen with the bulk as nitrate. Nitrate leaching from nursery containers is influenced by media temperature, plant root development, fertiliser type, method and efficiency of irrigation, and time since last fertiliser application.

Ammonium (NH_4^+) is less subject to leaching from soil or organic based media than nitrate (NO_3^-) because it can be held on cation exchange surfaces. Most soilless media have relatively little capacity to retain nitrate or other anions. Some plants grow better when ammonium, as well as nitrate, is supplied but high levels of ammonium will injure most plants. Organic fertilisers like blood and bone, blood meal, hoof and horn and various animal manures are considered relatively slow-release nitrogen sources. However, even these fertilisers produce the greatest flush of soluble nitrogen within the first few weeks after an application. This makes them useful for quickly greening up plants, but they are inferior to controlled release inorganic fertilisers for supplying nutrients over several months.

As most Australian soils are low in phosphorus and it is accepted

1

2

1 **Rose** – plants are stunted and have pale green foliage. The oldest leaves turn a uniform yellow or may clear from the veins. Purple spots may be present on leaf margins and between the veins. Oldest leaves senesce and are shed.

2 **Gerbera** – stunting and yellowing of the oldest leaves. Newer leaves are pale and small. Mature leaves develop a purple coloration spreading from the veins and from the margins of the blade. The pale green colour of foliage distinguishes this disorder from phosphorus deficiency which darkens leaves.

4

3

5

6

3 Chrysanthemum – plants are stunted with pale green foliage. The oldest leaves senesce prematurely becoming bright yellow and losing turgidity before being shed. Flower size and numbers are reduced by a severe deficiency.

4 Primula – seedlings have pale green, small leaves and grow very slowly. Oldest leaves may senesce prematurely.

5 Waratah – leaves are pale and small. (Healthy leaf at the top of the photo.) A few leaves at the very base of the plant senesce early, developing a brilliant red colour before falling.

6 Amaranthus – uniform pale green chlorosis of the older leaves followed by death of tissues near the tip of the blade. (Healthy leaf on right.)

7

8

9

7 Palm – leaflets on the oldest fronds develop a bright yellow chlorosis starting from the tips. This condition is followed by death and drying of the whole frond.

8 *Monstera deliciosa* – uniform bright yellow chlorosis of the oldest leaf and a pale green colour of younger leaves.

9 Syngonium – growth is reduced and the foliage is pale green. In variegated varieties, the leaf pattern in most mature leaves is poorly defined. The oldest leaves turn a uniform yellow and lose turgidity. These leaves later dry to a light brown papery condition.

10

10 Hydrangea – oldest leaves develop pale green, yellow and red colours while young leaves are a uniform pale green and small.

11 Lavender – plants are stunted and have pale green foliage. Older leaves turn a uniform yellow.

12 Black bean – plants are stunted with pale green leaves. Leaf size is greatly reduced.

11

12

13

14

15

13 Fiddle leaf fig – leaves begin to yellow from the base until the entire plant has a pale green colour. Oldest leaves senesce.

14 Ficus – uniform pale green to yellow coloration of all leaves.

15 *Picea omorika* – reduction in internode length and uniform pale green chlorosis of all leaves. (Photo: Dr D. Alt.)

PHOSPHORUS

M practice to apply liberal dressings of superphosphate or other p orus fertilisers to field-grown ornamental crops. For this reason, deficiencies of phosphorus tend to be less common than might be expected given the nature of our soil. Deficiencies do develop from time to time in crops grown in soilless media as when seedlings are grown without adequate fertiliser or where preplant fertilisers have been unevenly mixed through the growing medium. Symptoms can also appear in cold weather but these generally disappear when the soil temperature rises and roots become more active. Sometimes proteaceous plants which have been grown without phosphorus fertiliser to avoid phosphorus toxicity become mildly deficient.

Function Phosphorus is important for cell division and growth. It is needed for photosynthesis, sugar and starch formation, in energy transfer, and movement of carbohydrates within the plant. Plants vary in their need for phosphorus with many native plants having a low requirement as well as a low tolerance of phosphate. While most plants have around 0.25% phosphorus in the dry matter of their leaves, some including members of the Proteaceae grow well at levels down to 0.1%. Phosphorus is quite mobile within the plant and the older leaves are, therefore, the first to show symptoms of deficiency.

Symptoms A shortage of phosphorus will greatly reduce the growth of plant tops and roots. Flower numbers and size may also be reduced. Some plants produce proteoid roots in response to deficiency. This adaptation facilitates the absorption of phosphorus from the growing medium. Deficient seedlings may stop growing soon after emergence and cotyledons quickly become yellow and shrivel, before falling off. Deficient larger plants generally have stiff, erect leaves. These are a dull, dark green colour and it may be difficult to appreciate that a problem exists unless healthy plants are there for comparison. With increasing severity, new leaves become smaller and red and eventually yellow colours may appear on older leaves and on stems. This reddening is often more pronounced on the undersurface and along the veins of leaves. Generally, these senescent patterns are confined to one or two of the oldest leaves on a shoot, with other leaves remaining green. In contrast, leaf yellowing caused by nitrogen deficiency can be seen on leaves well up the shoot. In time, the lower leaves of phosphorus-deficient plants wither and die. These often remain attached to the shoot.

Corrective treatments

In soil Severe phosphorus deficiency is usually found in crops planted in new ground. Phosphorus does not readily leach from most soils, and residues left from previous fertiliser applications tend to build up. However, soil reserves will decline with cropping, especially under acid conditions, so some regular maintenance phosphorus is needed to maintain high yields. Application rates vary enormously (22–200 kg P/ha) depending on the crop, soil fertility, other soil properties, past fertiliser history and many other factors (see Table 4 on page 27).

Superphosphate, ammonium phosphates (MAP or DAP) and compound N–P–K fertilisers are the usual means of supplying phosphorus to soil grown crops. If used at high rates, poultry and animal manures can also contribute significantly to soil-phosphorus reserves. With soil-grown crops, all of the phosphorus fertiliser must be applied at or before planting to ensure good early root development and avoid crop setback. Surface applied phosphorus does not move easily through soil to roots, so side dressings are usually ineffective. Foliar sprays are also unsatisfactory because the amount which can be absorbed by leaves is small compared with crop requirement.

In soilless media Phosphorus is usually added during the preparation of container media as single superphosphate at rates of 0.7–2.4 kg/m^3. However, because phosphorus is mobile in soilless media, it can be supplied post planting either as a topdressing or in a liquid feed. The concentration of phosphorus in a liquid feed is normally in the range from 30 ppm to 60 ppm. The main phosphorus fertilisers are superphosphate and, in controlled release products or liquid feeds, ammonium or potassium phosphates. In the first week or two after potting up, phosphorus is leached from container media by water at a rate of around 9 g/m^3/month. Less phosphorus is lost if CRFs are used and when phosphorus is chemically fixed by soil or even sand present in the medium.

Phosphorus application rates for ornamental plants vary with the crop, growing medium, form of fertiliser, and method of application (see Table 4). A procedure for determining appropriate fertiliser rates is described on page 21. Foliage plants do not normally require basal superphosphate where a post planting phosphorus is applied. Phosphorus-sensitive plants may develop toxicity if phosphorus supply exceeds 10 g P/m^3/month which is less than one-third of the normal rate for nursery plants. These same plants may become deficient if the fertiliser program provides less than 5 g/m^3/month of phosphorus. Deficiencies can usually be corrected by topdressing pots with 0.5 g/L of single superphosphate.

1

3

2

4

1 Gerbera – plants are stunted and produce short-stemmed flowers. Older leaves develop a red coloration seen first on petioles and veins but eventually spreading over the blade. These older leaves may senesce prematurely. Younger leaves are a dark green colour. Flower stems are red.

2 Rose – plants are stunted and have sparse foliage. Older leaves initially develop a pale green chlorosis with the veins relatively green. With increasing deficiency, tissue near the leaf margin and between the veins reddens. Older leaves will often senesce prematurely, turning yellow before falling. Younger leaves are a dark green colour.

3 Cineraria – plants are stunted and leaves are small. Oldest leaves senesce early after first developing a uniform pale green chlorosis. Irregularly distributed areas of the blade may become water soaked and dry to a light brown colour. Younger leaves are a dark green colour.

4 Primula – phosphorus deficiency (left) causes poor seedling establishment. Plants are pale, stunted and have small leaves. Senescence is early in the oldest leaf which turns yellow before dying.

5 ***Protea neriifolia*** **cv. Pink Ice** – the deficient plant on the left is stunted and has smaller leaves than the healthy plant on the right. Leaves are abnormally dark and have a pink cast which is most developed in the youngest leaves. Stems are also red.

6 Cissus – plants are stunted and the foliage is a darker green than normal. The stems, petioles and the undersurface of leaves, particularly the veins, are purple.

7 ***Grevillea*** **cv. Royal Mantle** – deficient plants are less vigorous and develop a rust brown colour on stems and the undersurface of leaves. Compare deficient shoot on left with healthy shoot on right.

5

6

7

8

9

10

8 ***Syngonium*** – plants have dark green foliage which accentuates the variegated leaf pattern. Older leaves senesce prematurely turning a uniform pale yellow before drying to a papery consistency. The transition between chlorotic and green leaves is quite abrupt.

9 **Carnation** – growth is reduced. Older leaves turn a uniform yellow and contrast strongly with the dark blue–green colour of the other leaves. Pale brown blotches appear on older leaves and may eventually cover the entire leaf blade.

10 ***Picea omerika*** – stunting and premature shedding of needles. (Photo: Dr D. Alt.)

Potassium

Potassium deficiency is rarely the major nutrient limiting growth of plants in containers, probably because potassium fertilisers are routinely used. Symptoms of deficiency are more often seen in flowering crops planted on light soils. In soil, plants obtain potassium mainly from an exchangeable pool which is slowly replenished by the weathering of clay minerals. Sandy soils with their lower clay content are generally less well supplied with potassium than heavier textured soils and are most likely to become depleted of this element. Bark, sawdust and other organic materials supply some potassium to plants in soilless media, but these reserves are generally low compared with those of a soil so fertiliser additions are very important. Leaching losses of potassium are high due to the low cation exchange (retention) capacity of these media. Large amounts of potassium are removed from the soil in produce – flowers, leaves and stems – and these losses must be replaced to maintain fertility. Because flowers are mostly short-term crops with a high nutrient demand, the soil may be unable to meet their peak needs and application of potassium fertiliser is often necessary. The use of animal and bird manures as the major fertiliser source can also run down the potassium reserves of a soil because they are generally richer in nitrogen than potassium. Potassium deficiencies are more likely under hot, dry conditions.

Function Potassium is important for the formation of proteins, carbohydrates and fats, and for the functioning of chlorophyll and several enzymes. It is needed in cell division, for maintaining the balance of salts and water in plant cells, and for opening and closing the stomates, the tiny breathing pores on the undersurface of leaves. Concentrations of potassium are greatest in leaves, growing points, flowers and fruit. Potassium is highly mobile and moves freely within the plant to new tissues when needed. As a consequence, older leaves are the first to show deficiency symptoms.

Symptoms In most crops, potassium deficiency causes a marginal scorch on older leaves. This symptom usually commences as discrete spots in the leaf lamina between the major veins, which eventually coalesce forming light brown to almost black areas of dead tissue. Yellowing of the leaf margins and interveinal areas of older leaves often precedes scorching. Damaged leaves may curl downward or cup upward because the necrotic margin of the blade does not expand fully. Some potassium-deficient plants, including petunias, azaleas and pines, develop an iron-like chlorosis of new leaves as well as a scorch on older leaves. Potassium deficiency can reduce both the size and number of flowers on a plant and has been implicated in thin and brittle stem disorders. The incidence of calyx splitting and brittle stem in carnations exposed to low night temperatures ($<8^{\circ}C$) is significantly lower when potassium is adequately supplied. Potassium also seems important in conferring disease and frost resistance on plants. Yield and quality are reduced in deficient crops well before symptoms are apparent. Early warning of deficiency can be obtained by monitoring the potassium status of crops using leaf and soil analysis.

Corrective treatments Most flowering crops have a high demand for potassium, often taking up as much or more potassium than nitrogen. A large amount is absorbed into the flowers and so is removed from the farm at harvest. The short period (perhaps only two months or less) during which a flowering crop must absorb most of its potassium and other nutrients also means that potassium fertilisers are required, even in soils with adequate potassium for other crops. Absorption of potassium onto the soil exchange complex restricts movement through most soils, particularly those of a loamy or clayey nature, so leaching losses are minor except in the most sandy soils. The slow movement of potassium to roots from surface applications or side dressings means that the entire potassium requirement of the crop must be incorporated into the soil at or before planting. For the same reason, it is difficult to correct a deficiency in an annual field-grown crop, so prevention in future crops should be the aim. Foliar applications can be phytotoxic and, generally, will not provide enough potassium to save a deficient crop.

Potassium chloride (50% K) (muriate of potash) is the cheapest and the most commonly used potassium fertiliser. Potassium sulphate (42% K) contains less potassium and is more costly. Its use is only warranted where salinity is a major problem. Potassium nitrate (38% K, 13% N) contains nitrogen as well as potassium. It is expensive but a useful fertiliser particularly for liquid feeding. Mixed N–P–K fertilisers, available in a range of ratios of N:P:K to satisfy differing crop needs and soil conditions, are the most common means of maintaining balanced nutrition of soil-grown crops. Manures, composts and other organic soil amendments usually contain some potassium but tend to be richer in nitrogen and must be supplemented with inorganic potassium sources.

Potassium is present in large amounts in drainage water from nursery pots. Leaching losses are high because of over-irrigation, the low nutrient-holding capacity of most soilless media, and high inputs of soluble potassium fertilisers. In container production, potassium is normally provided to plants after planting as controlled release fertiliser or a liquid feed. Controlled release fertilisers are more resistant to leaching than uncoated soluble fertilisers. Around 80 g/m^3 of potassium is leached from container media each month from nurseries using overhead irrigation even when controlled release fertilisers are used. Losses are greatest in the first few weeks after potting up and in summer when irrigation is heaviest. Materials like zeolite, vermiculite and kaolite which have high cation exchange properties can reduce potassium leaching when added to media. Potassium concentrations in liquid feeds are normally in the range 100–200 ppm.

Excess When too much potassium is applied, the availability of calcium and magnesium for root uptake can be reduced sufficiently to cause a deficiency if either of these nutrients is in short supply.

2

1

3

4

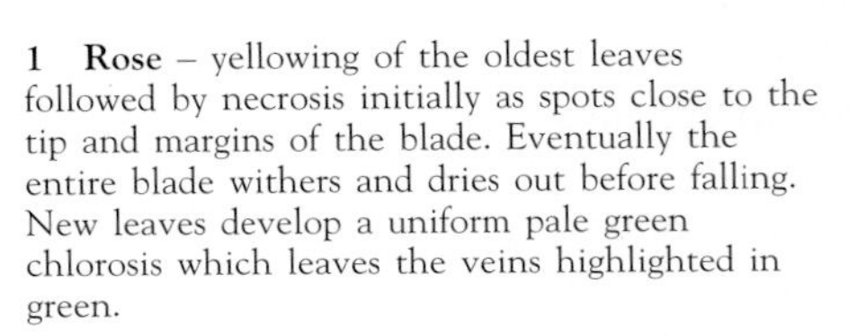

1 **Rose** – yellowing of the oldest leaves followed by necrosis initially as spots close to the tip and margins of the blade. Eventually the entire blade withers and dries out before falling. New leaves develop a uniform pale green chlorosis which leaves the veins highlighted in green.

2 **Rose** – an iron-like yellow–pink chlorosis of new leaves is a symptom of potassium deficiency in roses. Unexpanded leaflets are elongated and have wavy edges.

3 **Gerbera** – light brown necrotic spotting on the oldest leaves. Leaves are initially a dark purple–green colour. Necrosis begins between the veins and at the margins of the blade.

4 **Hollyhock** – photo showing symptom progression from mature to old leaves (left to right). Scorching of older leaves begins as a light green chlorosis which spreads from the margins back between the veins. Dead tissue turns light brown on drying.

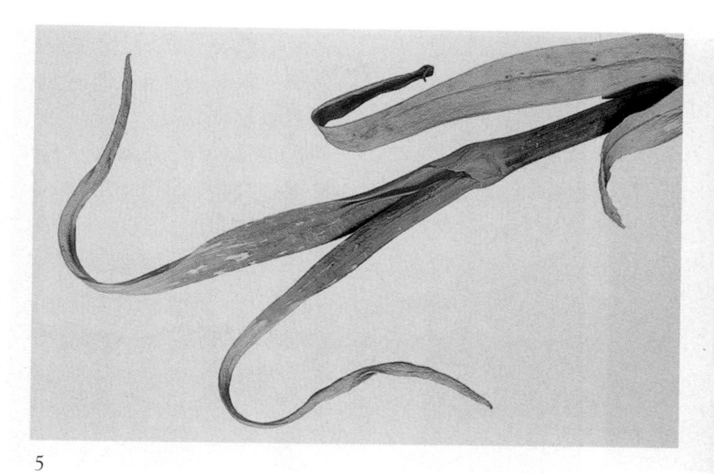

5

8

6

7

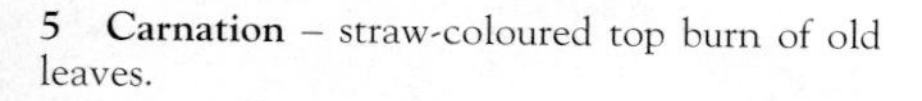

5 Carnation – straw-coloured top burn of old leaves.

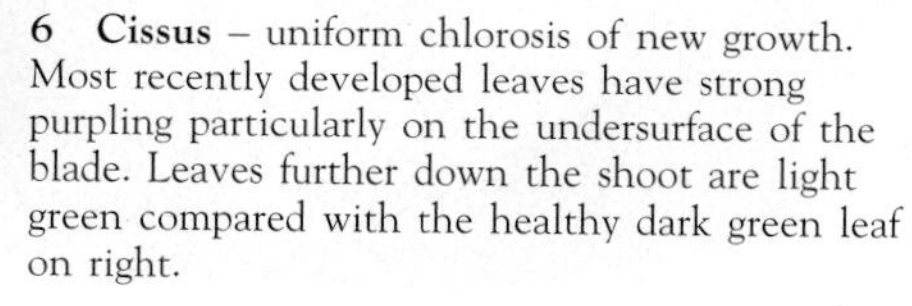

6 Cissus – uniform chlorosis of new growth. Most recently developed leaves have strong purpling particularly on the undersurface of the blade. Leaves further down the shoot are light green compared with the healthy dark green leaf on right.

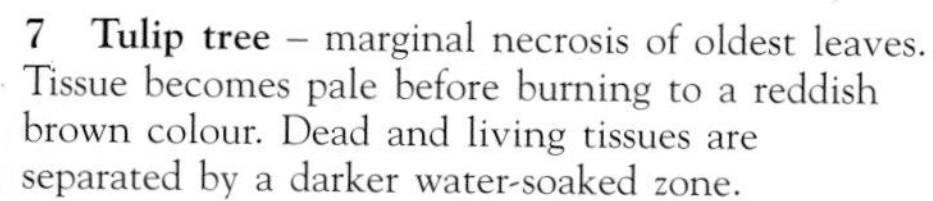

7 Tulip tree – marginal necrosis of oldest leaves. Tissue becomes pale before burning to a reddish brown colour. Dead and living tissues are separated by a darker water-soaked zone.

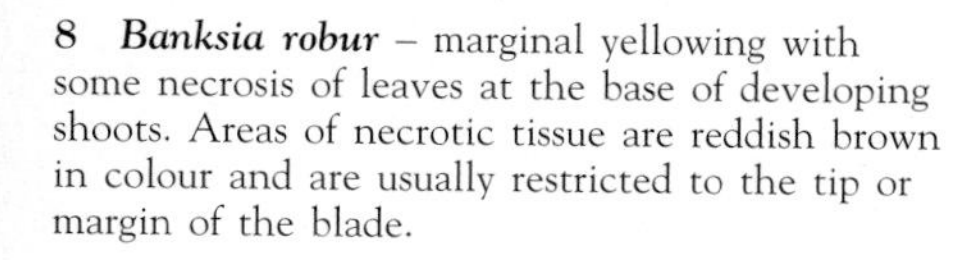

8 *Banksia robur* – marginal yellowing with some necrosis of leaves at the base of developing shoots. Areas of necrotic tissue are reddish brown in colour and are usually restricted to the tip or margin of the blade.

9

11

10

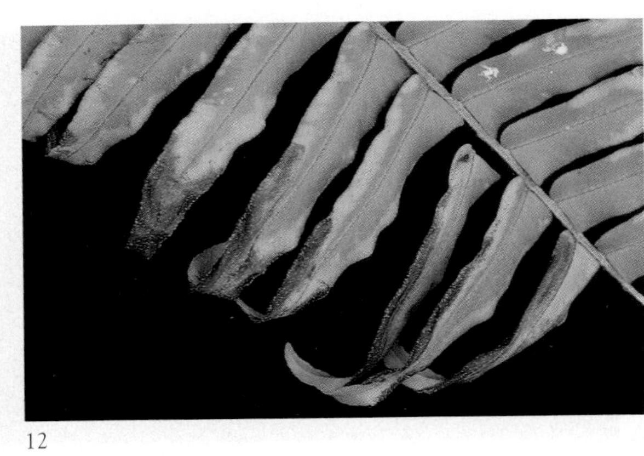

12

13

14

17

15

16

9 **Waratah** – reddish brown necrosis of the tip and margins of the oldest leaves which are pale green.

10 **Polyanthus** – interveinal yellowing of the oldest leaves. In severe cases, leaves at the base of the plant become uniformly yellow with the veins last to clear. The final stage is a brown burn on the margins of the leaf. Leaves are cupped downwards.

11 **African violet** – older leaves are light green in colour with purpling of the raised interveinal areas of the blade. Purpling is stronger on the youngest leaves which are also small.

12 **Boston fern** – brown burn on the margins of leaflets from basal fronds. Burn is preceded by a light green chlorosis.

13 ***Dracaena fragrans*** – light brown burn of the tips of old leaves. Symptom commences as small necrotic spots with a light halo between the veins. These areas of dead tissue enlarge and eventually coalesce and the leaf begins to roll inwards and shrivel from the tip. (Photo: Keith Bodman.)

14 **African violet** – potassium-deficient leaf showing purpling, light green border and moon-shaped marginal burn characteristic of this disorder.

15 **Rhododendron** – interveinal yellowing and burning of the leaf margins caused by potassium deficiency. This symptom will normally appear first in the oldest leaves, however, in this very severe example, leaves of all ages have been affected.

16 **Primula** – potassium-deficient seedlings are stunted and the oldest leaves have a pronounced yellowing between the veins. Yellowing spreads from the margins of the leaf towards the midrib and is associated with early senescence. In the final stages leaves dry to a light brown colour.

17 **Sweet pea** – yellowing and burning of the tip and margins of older leaves.

19

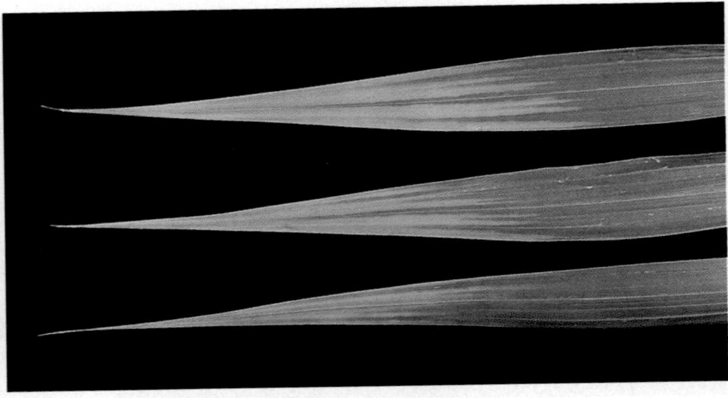

18

20

18 ***Howea forsteriana*, Kentia palm** – a pale green interveinal chlorosis followed by scorching commencing at the tips of the oldest leaflets.

19 ***Chrysalidocarpus lutescens*** – yellowing followed by scorching commencing at the tips and edges of leaflets on the oldest fronds. (Photo: Keith Bodman.)

20 ***Picea pungens* cv. Glauca** – potassium deficiency produces an iron-like chlorosis of the youngest leaves. (Photo: Dr D. Alt.)

CALCIUM

Calcium is less available to plants in leached, acid soils but deficiencies are not confined to these conditions. Symptoms appear when the supply of calcium to newly developing tissues is interrupted at a critical stage and this can happen even when soil reserves are adequate. Calcium deficiencies are more often the result of inadequate transport of calcium to tissues than of poor root uptake. Calcium moves in the transpiration steam, so is preferentially accumulated in tissues which transpire rapidly, like fully expanded mature leaves. In hot drying conditions, the supply of calcium to young leaves, growing points, root tips or fruit can become limiting because these organs transpire at lower rates relative to the older leaves. High humidity at night encourages calcium transport to susceptible tissues and reduces the risk of deficiency. However, when humidity is high by day and night, low transpiration flows in whole shoots can restrict calcium uptake and distribution leading to deficiency.

A plant's requirement for calcium increases with rapid growth. Deficiencies are, therefore, encouraged by high light, heavy fertiliser use, warm weather and moist soil conditions. Over-use of ammonium or potassium fertilisers can competitively reduce the availability of calcium for root uptake. Uptake is also restricted by conditions which impede root growth including poor aeration due to compaction, high media temperature, salinity, drought and disease.

In nurseries relying on overhead irrigation, around 86 g/m^3/month of calcium is leached from container media even though lime and dolomite are relatively insoluble in water. Calcium is naturally present in many irrigation waters and, where calcium concentrations are above 100 ppm no additional supplies are needed to grow plants.

Function Calcium is an important constituent of cell walls and membranes and, when it is in short supply, these membranes become leaky and cell division is disrupted causing abnormalities in the growing points and root tips. Calcium appears to be important in protecting plant cells from toxins, in slowing the aging of plant tissues, and conferring tolerance of certain diseases. Calcium is found at percentage levels in plant tissues, but concentrations associated with deficiency are poorly defined. Calcium has poor mobility within the plant, so deficiency symptoms appear first in young tissues.

Symptoms Calcium deficiency symptoms appear on rapidly growing tissues, such as the shoot tip, expanding leaves, flowers, fruit and roots. They include necrosis and distortion of young leaves commencing either as a water soak or a drying of marginal tissues, wilting of flower stems, premature flower opening and root damage. When deficiency is severe, the growing points and flower buds may die leading to multiple branching and stunting of the plant. In carnations, poinsettias and chrysanthemums, calcium deficiency retards plant growth and causes a pale brown scorch and constriction of the tips or margins of expanding leaves. Deficient palms produce stunted, distorted new leaves which stick in the bud and, consequently, do not open normally. Leaflets are often necrotic with only the base of the petiole remaining alive and, when severely deficient, the growing point of the palm dies. Calcium-deficient plants are susceptible to infestation by pests and diseases. Deficiencies are difficult to diagnose using conventional plant and soil testing methods and much reliance is placed on symptoms.

Corrective treatments With few exceptions, most crops grow well in media which are only slightly acid, with a pH about 5.5–6.0 (1:5 $CaCl_2$) or 6.0–6.5 if the pH is measured in water. Some acid-loving plants, like azaleas and many native proteaceous plants, are adapted to lower pHs. Where the growing medium is more acid than desired, lime (calcium carbonate) or dolomite (calcium and magnesium carbonate) can be added to raise the pH and supply calcium. Rates of lime used in preparation of container media generally range from 3–15 kg/m^3 depending on the quality of the lime, the buffering capacity of the mix and the need for pH adjustment. Superphosphate and gypsum are other important sources of calcium in these mixes. Gypsum is generally used when no pH change is needed. Around 17 kg/m^3 of gypsum will provide adequate plant calcium. Although liming increases the supply of calcium to the roots, it cannot prevent calcium deficiency if the environmental conditions favour its development. Some modification of the environment is possible. Covered crops should be well ventilated during the day to reduce humidity and increase transpiration. In hot weather, sprinkler irrigation can be used to cool the crop and reduce transpiration. Misting of susceptible crops at night if humidity is low may reduce calcium deficiency. However, this treatment will increase the incidence of fungal and bacterial diseases if the foliage remains wet for long periods.

Calcium sprays can be used to prevent deficiencies in ornamental crops but, to be effective, they must contact the susceptible tissue and be applied quite frequently during periods of active growth. Under these conditions a weekly spray of calcium nitrate (800 g/100 L) can reduce the risk of deficiency. Four sprays of calcium chloride (80 g/100 L) are recommended to control marginal bract necrosis in poinsettias. Other steps which may be taken to prevent deficiency include growing tolerant varieties, managing fertiliser use to reduce growth rates and to avoid antagonism from competing ions like K^+ and NH_4^+, and avoiding waterlogging, moisture stress and salinity.

1

2

3

4

1 **Chrysanthemum** – the youngest leaves develop necrotic tips giving the blade a squared off appearance.

2 **Stocks** – necrosis of tips of the youngest leaves.

3 **Petunia** – an interveinal mottling and necrosis of the younger leaves. Tissues dry to a light brown colour with papery consistency. (Photo: Kevin Handreck.)

4 **Carnations** – death and bleaching of the tips of the youngest leaves

5

7

6

8

9

10

5 Lisianthus – calcium deficiency causes scorching of the tips of the youngest leaves and may result in the death of the growing point.

6 Tuber rose – reddish brown bruise-like markings on the tips and edges of sepals caused by calcium deficiency.

7 Sinningia – necrosis of the growing point and flower buds caused by calcium deficiency. (Photo Dr D. Alt.)

8 Birdsnest fern – bruise-like injury to the tip and margins of expanding leaves causing constriction and deformation of the leaf blade. Areas of necrosis are light green in colour and may dry to a light brown.

9 Poinsettia – a marginal necrosis of the floral bracts caused by calcium deficiency. (Photo: Dr A. Wissemeier.)

10 Tulip – bent neck. A condition where the flower stem collapses and folds over is a symptom of calcium deficiency. (Photo: Dr D. Alt.)

MAGNESIUM

Magnesium deficiency is one of the more common disorders of ornamental plants. Most light-textured soils, both coastal and inland, will become deficient if dolomite or magnesite is not used regularly as part of the liming program to counter acidification. The risk of magnesium deficiency is also increased if use of potassium fertilisers or of agricultural lime has been excessive. This is because potassium and calcium are able to displace magnesium from cation exchange sites in the soil, aiding its movement and loss beyond the root zone.

Most materials used to produce soilless media are inherently low in magnesium so, to avoid deficiency, supplements must be added in preparation or in the regular fertiliser program. Dolomite and magnesium sulphate are the usual preplant sources, and magnesium sulphate or magnesium nitrate if magnesium is needed after planting. Some controlled release fertilisers also contain magnesium. The low cation exchange capacities of most soilless media allow relatively high rates of magnesium leaching. Losses of around 31 g Mg/m^3 each month are normal in Australian nurseries. Around 30 ppm of magnesium is required in constant feed solutions for crops grown in inert media such as rockwool and perlite.

Function Magnesium is a component of chlorophyll, the vital green pigment which enables plants to synthesise sugars and starches from atmospheric carbon dioxide. It also activates enzymes involved in energy transfer processes. Magnesium is mobile within the plant, moving from older to newer tissues at times of shortage. As a consequence, the oldest leaves are the first to show deficiency symptoms. For most plants, a leaf magnesium concentration above 0.25% is normal but concentrations as low as 0.1% appear adequate for many proteaceous plants.

Symptoms Most magnesium-deficient crops develop a characteristic bright yellow chlorosis in older leaves. The yellowing begins at the tip and margins of the blade and spreads inward between the veins towards the midrib, sometimes leaving a triangular green area near the leaf base. In palms, the entire leaf may become yellow. Plants with parallel leaf venation may develop a green tram track symptom as the interveinal areas turn yellow. Reddish purple blotches are produced on chlorotic leaves of some crops such as roses and azaleas. Severe deficiencies usually cause a rusty brown burn in the yellow regions of the blade beginning at the tip and margins of the older leaves, which generally shed prematurely. With some plants, a few leaves at the very base of the shoot may fail to show symptoms. Magnesium deficiency symptoms can be confused with those of potassium deficiency or manganese deficiency. The most reliable way of distinguishing between them is to conduct a leaf analysis, however, some differences in symptoms do exist. Manganese symptoms usually appear on recently matured and younger leaves whereas the symptoms of magnesium deficiency are found on the oldest leaves. Burning, which is characteristic of severe magnesium deficiency, is almost never caused by manganese deficiency. Although there are significant exceptions, as a general rule, magnesium deficiency is typified by chlorosis and potassium deficiencies by necrosis or scorching.

Corrective treatments On deficient soil, 300 kg magnesite or 800 kg dolomite per hectare applied and worked in prior to planting should prevent magnesium deficiency for several years. Dolomite may also be used in soilless container media as a magnesium source as well as a liming agent. A rate of 5 kg/m^3 of good quality dolomite is usually adequate in a peat-based medium. Fortnightly foliar sprays of 2% (2 kg/100 L) magnesium sulphate (Epsom salts) at high volume (500–1000 L/ha) may be used to control a deficiency in a growing crop. Magnesium can also be applied as a soil drench to deficient plants at a rate of 240 g/100 L (no more than 50 g/m^2) of magnesium sulphate.

1

1, 2 and 3 Rose – magnesium deficiency causes a light green chlorosis of the margins and interveinal tissues of older leaves. Typically a V-shaped region of healthy green tissue is left at the base of the leaflet. In severe cases, chlorotic areas turn yellow and may burn (**2**). In some varieties, purple green blotches appear on the older leaves prior to chlorosis (**3**).

2

3

4 Chrysanthemum – yellow chlorosis of older leaves commencing at the margins of the leaf blade and progressing inwards between the major veins.

5 Gerbera – mottling of the older leaves together with some purpling apparent on the upper surface of the blade especially the raised portions between the veins. Affected leaves are often brittle.

6 Primula – stunting and yellowing are the main symptoms of magnesium deficiency in seedlings (right).

4

5

6

7

8

9

7 Hollyhock – magnesium deficiency causes severe interveinal yellowing of older leaves leaving a net-like pattern of green veins.

8 Daphne – a uniform yellow chlorosis which starts at the tips and margins of the older leaves and progresses back towards the base of the leaf, leaving a V-shaped area of healthy green tissue.

9 Umbrella tree – individual leaflets develop a pale green interveinal chlorosis which commences at the leaf margin and moves towards the mid vein.

10 Philodendron – the first symptom is an interveinal yellowing starting at the margins of the oldest leaves (left). In its severest form (right) the veins are clear, and browning and scorching is apparent on the tips and around the central veins.

11 Hibiscus – magnesium deficiency produces symptoms on the older leaves.

12 Japanese sacred bamboo – marginal and interveinal chlorosis of the older leaves resulting from magnesium deficiency.

13 *Banksia robur* – pale green interveinal chlorosis followed by necrosis. Symptom is most developed between the margin and the vein. Healthy leaf on left for comparison.

14 Banksia – light green chlorosis of older leaves starting at the tip and margins of the leaf and extending back towards the base. A wedge-shaped area of healthy green tissue may be present near the petiole.

15 and 16 *Phoenix roebelenii* – yellowing and scorching of the oldest fronds. The condition spreads from the tips towards the base of leaflets leaving an area of healthy green tissue (**16**). (Photos: Keith Bodman.)

10

11

12

13

14

15

16

SULPHUR

Field-grown ornamental crops rarely develop sulphur deficiency because adequate sulphur can usually be obtained from soil organic matter, manure, and fertilisers such as superphosphate, ammonium sulphate, potassium sulphate and gypsum. Traces of sulphur are also present in domestic water and rainwater, and in horticultural sprays (micronutrients and fungicides).

Plants in soilless media which have received only high analysis fertilisers and have been irrigated with water containing less than 25 ppm sulphur can develop sulphur deficiency. Most soilless media supply very little sulphur and have little capacity to retain sulphates against leaching due to low anion exchange properties. Most of the soluble sulphur in fertilisers, as well as the less soluble sulphur from superphosphate and gypsum, is leached from containers in irrigation water within the first 7 weeks of growth. Around 75 g S/m^3 is lost from nursery container media in a month under Australian conditions.

Function Sulphur is a constituent of protein, and of amino acids and co-enzymes, and is involved in many critical plant processes. It occurs at percentage levels in plant tissue, and 0.1–0.2% of sulphur in leaves is usually adequate. Sulphur has poor mobility within the plant and, when supply is grossly inadequate, symptoms usually show first in the youngest leaves.

Symptoms The youngest leaves of severely deficient plants become uniformly chlorotic, contrasting with the dark green colour of neighbouring older leaves. This symptom can be difficult to distinguish from iron deficiency in some crops. However, unlike iron chlorosis, the veins of sulphur-deficient leaves do not remain green and the leaf blade is a dull milky yellow. If the deficiency persists, growth of tops slows and newer leaves become smaller. Eventually, leaves of all ages become paler and the most chlorotic leaves can burn from the tips.

With a mild sulphur deficiency, young leaves are often green while all mature leaves become pale. This is because sufficient sulphur can be remobilised from older tissues to satisfy the needs of the new leaves. In time, the oldest leaves turn yellow and senesce prematurely. The unpatterned chlorosis is then easily confused with that of nitrogen deficiency.

Corrective treatments Container-grown plants obtain much of their sulphur from superphosphate and gypsum incorporated into the growing medium before planting. However, if there is less than 30 ppm sulphur in the water supply, an additional source may be needed to compensate for losses through leaching. Ammonium, potassium or magnesium sulphates may be used in liquid feed at rates up to 100 ppm of sulphur. No more than 60 ppm sulphur should be used for continuously fed plants. Controlled release fertilisers are a useful supplementary source of sulphur but may be inadequate as the sole source for container-grown plants. Sulphur is potentially limiting in a soilless medium when there is less than 6 ppm sulphur in a 1:1.5 v/v water extract or less than 16 ppm in the drainage water.

1

1 **Rose** – sulphur deficiency causes a uniform yellowing of the younger leaves with the veins remaining green. Healthy leaf (right) is given for comparison.

2 **Gerbera** – plants are stunted and produce small narrow leaves. New leaves exhibit an interveinal chlorosis.

3 **Hollyhock** – a uniform chlorosis of the younger leaves resembling iron deficiency.

3

4

5

6

7

4 Cineraria – sulphur-deficient plant (right) has pale green foliage. Oldest shaded leaves are yellowing.

5 Petunia – yellow–green chlorosis especially of the younger leaves. (Photo: Dr D. Alt.)

6 *Banksia robur* – uniform yellow chlorosis of the youngest leaves. Mature leaf on right has mild interveinal chlorosis.

7 Waratah – sulfur-deficient young leaves on right are pale green compared with the oldest leaves (left). Light brown scorch can develop as the disorder progresses.

BORON

For optimum growth, all plants need an uninterrupted supply of boron. Plant requirements for boron are relatively low but tend to be higher for dicotyledons (broadleaved plants) than for monocotyledons (grasses, lilies, etc.). Crops susceptible to boron deficiency include carnations, azaleas, cyclamen, gardenias, geraniums, gladioli, roses, gerberas, chrysanthemums, begonias, stocks, palms and pines. Soils differ in their capacity to supply boron; light-textured soils (for example, those derived from granites or sandstone) that are low in organic matter and in high rainfall areas are usually more prone to deficiency than heavier textured soils in drier areas. Soils derived from marine sediments often have the highest reserves of boron. The availability of boron for plant uptake is pH-dependent and deficiencies often occur after heavy liming. Unlike other trace elements, boron is easily leached from the soil. Plant uptake is also reduced in dry soil, and deficiencies mostly occur when a dry spell follows a long period of wet weather. Plants are most susceptible to deficiency in warm sunny conditions and least in overcast, cool weather. This occurs, in part, because plants require more boron when they are growing rapidly. The strong influence of weather conditions on boron availability to plants explains why deficiencies can suddenly reappear even though management has not altered.

Function Boron has a role in regulating cell development in growing points and in transporting sugars within the plant. Pollen production and pollen tube development are both impaired when it is deficient, reducing numbers of viable seeds and fruit development. There is little remobilisation of boron from older leaves in most plants and so deficiencies show first in the youngest tissues – buds, leaves, flowers and fruit – even after a relatively short interruption in supply. Most dicots require around 30 ppm of boron in dry matter of leaves to grow normally whereas monocots (grasses) require only 5–15 ppm.

Symptoms Young rapidly growing leaves develop a water-soaked necrosis beginning at the tip of the blade. This causes a pinching of the leaf as it expands and is similar to the symptom for calcium deficiency. The foliage and stem tissues of boron-deficient plants are brittle and prone to cracking and splitting. Flowering may be delayed and a high proportion of buds abort. Flowers are often small with deformed and abnormally coloured petals. Closely spaced transverse cracks or corkiness may develop on the surfaces of petioles, stems, and the midribs of leaves. This may be accompanied by bronzing of petioles and staining of vascular tissue. Internodes shorten and the terminal shoot will often die leading to the development of multiple side shoots, a condition known as 'witches broom'. Boron-deficient carnations have a higher incidence of calyx splitting and bud abortion, and those buds that develop have a pronounced style and fewer petals. On lower leaves, red and necrotic areas appear near the mid veins. Multiple side branching gives a bunchy appearance to the affected carnation plant.

Boron deficiency causes shortening and splitting of the flower stalk in gerberas, and a petal deformity called 'quilling' in chrysanthemums. New leaves of deficient palms are chlorotic and malformed. As with calcium deficiency, leaf expansion is restricted and, in severe cases, the growing point dies. Boron deficiency in gladioli causes curving of leaf tips and V-shaped breaks near the inner base of large leaves. Leaf edges are thin, colourless and cracked.

Symptoms on leaves are less reliable for diagnosis than those on buds and flowers. They include distortion of newer leaves in azaleas, roses, begonias, gardenias and stocks; chlorosis and necrosis in chrysanthemum, cyclamen and gardenia; and leathery leaves in chrysanthemums and geraniums.

Corrective treatments A preventative soil application of 1–2 kg/ha of boron (10–20 kg/ha borax) at or prior to sowing is preferable as damage caused by boron deficiency is usually serious and permanent in an annual crop once symptoms appear. Soil treatments are normally effective for more than one year but longevity is influenced by soil type and weather conditions. A foliar spray of 500 g/ha of boron (2.5 kg/ha Solubor at a strength of 0.1–0.2% w/v) used before any symptoms develop may prevent serious deficiency and is safe for most situations but a test treatment is advisable to ensure that the rate is not phytotoxic. A foliar application will not protect following crops, so spray treatments must be repeated every season. Nutrient solutions for crops in soilless media (hydroponics) normally contain 0.2–2.0 ppm of boron. DTPA extractable boron in the range of 0.2–0.65 ppm is considered optimal for plants in soilless potting media (Australian Potting Mix Standard).

1

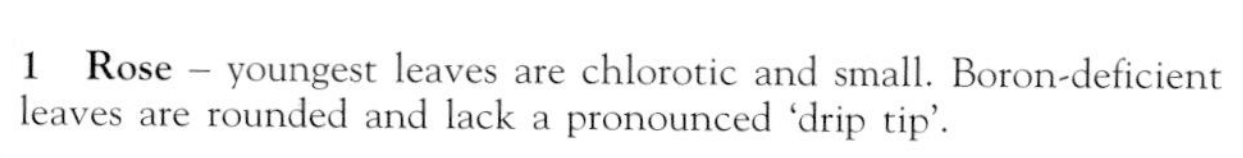

1 **Rose** – youngest leaves are chlorotic and small. Boron-deficient leaves are rounded and lack a pronounced 'drip tip'.

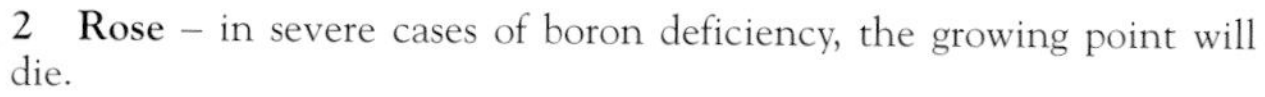

2 **Rose** – in severe cases of boron deficiency, the growing point will die.

2

4

3

5

6

3 **Gerbera** – stunting and distortion of leaves and flowers caused by boron deficiency.

4 **Carnation** – multiple branching of shoots and abortion of the floral bud (far left), and flower deformities are symptoms of boron deficiency.

5 **Gerbera** – examples of leaf distortion caused by boron deficiency.

6 **Carnation** – irregular development of the flower, failure of petals to emerge from the bud and bud abortion are the major floral symptoms of boron deficiency.

8

7 Carnation – cracking and corkiness around the stem nodes and abnormal production of auxillary shoots are the major vegetative symptoms of boron deficiency.

8 Hollyhock – distortion of younger leaves caused by boron deficiency. Leaves develop a marginal necrosis at an early stage and do not expand completely. Longitudinal cracks may appear on leaf petioles.

9 Hollyhock – tissue between the major veins of mature leaves may collapse and dry to a light tan colour.

9

7

10

11

12

13

14

15

13 Boronia – multiple branching and death of the primary growing point caused by boron deficiency.

14 ***Howea forsteriana*** – scorching of unfurled fronds and death of the growing point are the primary symptoms of boron deficiency in the Kentia palm.

15 Pine – death of the growing point and reduced needle length caused by boron deficiency. (Photo: Dr D. Alt.)

10 Sinningia – necrosis of the growing point and the margins and interveinal regions of younger leaves caused by boron deficiency. (Photo: Dr D. Alt.)

11 Cissus – leaves from a boron-deficient plant are pale and mottled compared with healthy leaves (right of photo). Younger leaves are small and distorted.

12 ***Heliconia caribea*** – water soaked markings develop in the interveinal region of young leaves under boron deficiency.

IRON

Iron deficiency, sometimes called 'lime-induced chlorosis' because it is common in crops on calcareous soils, is probably the best known trace element disorder of ornamental plants. It develops when there is too little soluble iron in the growing medium or when the iron which has been absorbed by the plant is rendered unavailable for developing tissues.

Iron availability for root uptake is reduced at high pH (pH>7.5) and, therefore, deficiencies are often found in soils which have been over-limed, or which have free lime concretions near the surface, or which have been contaminated with brickies lime on building sites. Alkaline soil conditions and high concentrations of calcium carbonate or phosphorus, in media or water, cause deficiency by reducing the solubility of iron in the soil solution. Loss of root function due to low soil temperature, high levels of salt or ammonia in the medium, waterlogging, soil compaction, mechanical root pruning during transplanting, nematodes and root rots, can also induce iron chlorosis symptoms in rapidly growing plants.

Even after iron has been absorbed by plants nutrient imbalances can reduce its physiological usefulness. Plants suffering from toxicities of phosphorus, copper, zinc and manganese will generally have symptoms of iron deficiency even when leaf iron levels are high. Furthermore, the recovery of these plants can be assisted by supplying additional iron. This is the case with phosphorus toxicity in some proteaceous plants. In many plants iron chlorosis symptoms are produced in response to severe potassium deficiency. Many plants, including palms, grevilleas and other native proteaceous plants, develop iron-like deficiency symptoms during the spring growth flush. This is probably due to low root activity in cold soil as the symptoms usually disappear as the weather warms up.

Plant species and cultivars differ in their sensitivity to iron deficiency. This appears to be due to differences in the ability to obtain iron from the growing medium rather than to differences in tissue requirement. Iron efficient species improve uptake by excreting substances from roots which increase the solubility of iron in the soil solution. Plants that are unable to do this (Table 7) grow best in a slightly acid medium (pH 5.5-6.0 1:1.5 water) which has been amended with iron. Azaleas, camellias, roses, rhododendron and many proteaceous plants are in this group.

Function Plants require iron to produce chlorophyll and to activate several enzymes including those involved in the oxidation/reduction processes of photosynthesis and respiration. Iron concentrations of 50–100 ppm in leaves are satisfactory for most crops. Higher levels are occasionally found in washed deficient leaves, which indicates that iron can exist in physiologically unavailable forms within plant tissues. Therefore, while a low iron value in a leaf analysis test indicates that iron is deficient, a normal or high iron concentration does not rule out the possibility of deficiency. Iron is relatively immobile in plant tissues and so the first symptoms of deficiency appear in the youngest leaves.

Table 7 Ornamental plants susceptible to iron deficiency

Abeliophyllum
Acacia
Acer
Adenocarpus
Aegle
Akebia
Amelanchier
Arbutus
Arctotis
Astilbe
Azalea
Banksia
Begonia
Berberis and Mahonia
Boronia
Callicarpa
Callistemon
Calycanthus
Camellia
Cassinia
Castanea
Catalpa
Ceanothus
Chaenomeles
Chimonanthus
Chrysanthemum
Cineraria
Citrus
Clematis
Clethra
Crowea
Cyclamen
Cypripedium
Cytisus
Dampiera
Daphne
Delphinium
Desfontainea
Escallonia
Eucalyptus
Eucryphia
Exochorda
Fraxinus
Fern
Fir
Garrya
Gerbera
Gladiolus
Gloxinia
Grevillea
Hemerocallis
Hibiscus
Hydrangea
Hypericum
Hypocalymma
Ipomoea
Kerria
Lapageria
Leontopodium
Leptospermum
Liquidambar
Lonicera
Lupin
Magnolia
Mesembryanthemum
Michaelmas Daisy
Oak
Osmanthus
Oxypetalum
Peony
Pansy
Piggyback plant
Phlox
Pelargonium
Petunia
Pinus
Populus
Potentilla
Poterium
Primula
Prostanthera
Prunus
Pyracantha
Pyrethrum
Quercus
Rosa
Rhododendron
Salvia
Soldanella
Spiraea
Stephanandra
Sweet pea
Telopea
Thuja
Tilia
Tropaeolum
Tsuga
Viburnum
Vitis
Weigela
Wisteria

Symptoms In the early stages of the disorder, the youngest leaves develop a uniform light green chlorosis with only the veins remaining dark green. This produces a distinctive net-like vein pattern. If the condition is severe and persistent, the chlorosis becomes more yellow or even white and the veins also lose their green colour. These chlorotic areas are relatively sensitive to burns from sprays or sunlight. Because iron does not move freely within the plant once it has been assimilated, older leaves often remain dark green while new emerging leaves are quite chlorotic. Iron deficiency also causes browning and stunting of roots, root elongation ceases, and additional lateral roots and root hairs are produced. The clearest iron deficiency symptoms are usually expressed when plants are growing rapidly, under high light conditions.

Corrective treatments Applying iron to soil gives unreliable results and is expensive when chelates are used for large areas. A trial application of 1–2 g of iron sulphate or iron chelate per metre of row may be justified for some high value glasshouse crops. Where the soil or growing medium is too alkaline, it can be acidified with iron sulphate, aluminium sulphate, elemental sulphur or a dilute solution of phosphoric acid. The appropriate rate and frequency of application depends on the buffering capacity of the soil or growing medium and should be established in a small trial before being tried on large numbers of plants. Alternatively, the growing medium can be watered with an iron sulphate (5 g/L) solution every week until the desired pH has been obtained. For chlorotic azaleas growing in soil of pH above 6.5, 25 g iron sulphate per square metre, watered into the soil will lower the pH and supply iron. The treatment may need repeating every 6–8 weeks until the foliage regreens or the pH falls below 6.0. Foliar sprays of 0.05% iron chelate (Fe EDTA) at high volume (500–1000 L/ha) at fortnightly intervals may give temporary control until the soil treatment takes effect. Iron sulphate (5–10 g/L) may also be used for foliar applications but generally this treatment gives less uniform regreening of leaves than the chelate.

Long-term control of iron deficiency in field-grown crops can be achieved by managing drainage and irrigation to avoid waterlogging, selection of more tolerant crops or cultivars, and using acidifying fertilisers such as ammonium sulphate. Some crops given just nitrate nitrogen become more susceptible to iron deficiency, so as a general rule an ammonium source of nitrogen should also be used. With hydroponic crops, the pH of the nutrient solution and media, should be kept at about 6.0–6.5 and the solution should be changed or adjusted regularly to avoid nutrient imbalances developing. Iron chelates are preferred for hydroponics as they are more likely to remain in solution than iron sulphate. Iron deficiency can be prevented by adding iron sulphate (250–750 g/m^3) or iron chelate (15–40 g/m^3 EDTA or 30–80 g/m^3 EDDHA) to a soilless potting medium during preparation, ensuring that its pH is slightly acid and that phosphorus application is not excessive. Higher rates of iron sulphate (1.5–2 kg/m^3) should be used with phosphorus-sensitive plants and with media based on pine bark or sawdust which have a relatively high capacity to immobilise iron. Iron may be supplied to established crops in the irrigation water as iron sulphate (0.25 g/L) or Fe EDTA (0.05 g/L). The Australian Standard for potting mixes specifies a minimum of 35 ppm of DTPA extractable iron for optimum plant growth.

3

1

2

4

5

6

7

1 **Rose** – uniform pale green chlorosis of the youngest leaves and a characteristic net-like pattern of green veins.

2 and 3 **Chrysanthemum** – uniform yellow chlorosis of youngest leaves (2). Iron deficiency causes death of roots which appear short and thickened, and brown (3).

4 **Lithianthus** – the iron-deficient plant on left is stunted and pale green in colour.

5 **Hydrangea** – uniform pale green chlorosis of younger leaves with characteristic net-like pattern of veins.

6 **Cineraria** – chlorosis of youngest leaves.

7 **Primula** – chlorosis of youngest leaves.

8 **Petunia** – chlorosis of youngest leaves.

9 **Pittosporum** – uniform pale green chlorosis of leaves with a border of healthy green tissue around the main vein.

10 **Azalea** – chlorosis of youngest leaves.

8

9

10

11

12

13

11 Syngonium – pale green chlorosis of the youngest leaf.

12 Ficus – chlorosis of youngest leaves.

13 Christmas bush – the dark green blotches seen on the chlorotic leaf (top centre) are a response to applied iron solution.

14 *Grevillea* cv. Royal Mantle – pale green chlorosis of the youngest leaves.

14

15

16

17

15 ***Banksia robur*** – when iron deficiency is severe, the veins also lose their green colour and the new leaves appear bleached. Net-like pattern can still be seen in older leaves.

16 Geraldton wax – distinctive tip yellowing symptom caused by iron deficiency.

17 Phoenix palm – young iron-deficient frond (right) is chlorotic.

MANGANESE

Leaf symptoms of manganese deficiency are not uncommon in ornamental crops. Like iron, manganese is less available for plant uptake when the growing conditions are alkaline. Heavy liming, especially of light, sandy, poorly buffered soils, can produce mild to moderate leaf symptoms in sensitive crops. Deficiencies are also caused by excessive biological immobilisation of manganese which can occur when high rates of organic matter (sewage sludge, crop residues or animal manures) are incorporated into soils. This biological pool of manganese can also be released suddenly, causing toxicity, when soil or medium is sterilised.

Transient deficiencies indicated by mild mottling in leaves often develop during cool growing conditions when root activity is low. These symptoms will generally disappear as the weather warms up and do not respond to soil treatment. Injuries to roots caused by poor aeration, mechanical damage during transplanting, nematode activity and diseases can also cause manganese deficiency.

Plant manganese status appears to be linked in some crops to disease resistance. The relationship varies according to the disease and the host species, however, generally disease severity is reduced as manganese supply is increased. This means that leaf manganese levels are usually lower in diseased plants.

Function Manganese is needed for chlorophyll formation, photosynthesis, respiration, nitrate assimilation, and the activity of several enzymes. Foliar manganese concentrations of 30–100 ppm in dry matter are satisfactory for most crops. Manganese is only moderately mobile in plant tissues and so symptoms appear first in younger leaves.

Symptoms The uppermost leaves on a shoot develop an interveinal chlorosis with major veins remaining green while the tissue between becomes progressively more yellow. Leaf symptoms of manganese and iron deficiency are similar and can be confused. The main differences are that manganese-deficient leaves have a greenish yellow chlorosis with a narrow strip of green tissue bordering the major veins, whereas in iron-deficient leaves the chlorosis is whitish green and only the veins remain green. Manganese-deficient plants produce normal-sized leaves, even under extreme deficiency (cf. zinc deficiency) but they may accumulate abnormally high levels of nitrate. This can be measured by a quick sap test.

Some manganese-deficient plants in the genera *Spathiphyllum*, *Aglaonema*, *Rhektophyllum*, *Monstera*, *Schismatoglottis*, *Caladium*, *Philodendron*, *Maranta*, *Calthea* and *Heliconia* develop an interveinal necrosis which causes the blade to split from the mid vein almost to the margin. This unusual symptom is known as 'skeleton leaf'. Deficient plants normally have less than 20 ppm manganese in leaves and diagnosis can be readily confirmed using leaf analysis. Palms may develop a zig-zag distortion ('fizzle top') of the young fronds which are pale yellow and stunted. The lower leaflets of the youngest fronds may eventually die.

Corrective treatments The pH of the soil or growing medium should be slightly acid (less than 6.5 1:1.5 water). Avoid over-liming. Manganese sulphate applied as a series of foliar sprays (1–2 g/L), as a side dressing (9 kg/ha), or a root drench (5 g/m^2 or 2.4 g/10 L) to soilless media, will correct a deficient crop. A preplant application of manganese sulphate at 10 kg/m^3 should satisfy the requirements of most plants in soilless media. The Australian Standard for potting mixes identifies the acceptable range for manganese as 1–15 ppm (DTPA extractable).

1

2

1 **Rose** – uniform pale green interveinal chlorosis of upper to middle leaves on a shoot. Major veins are bordered by healthy green tissue.

2 **Rose** – in some varieties even the smaller veins remain green giving a more reticulated pattern of chlorosis.

3

4

5

3 **Gerbera** – strong interveinal chlorosis of the younger leaves including the most recently mature leaf. Shortened flower stems.

4 **Chrysanthemum** – interveinal chlorosis of younger leaves. Healthy leaf on left for comparison.

5 **Hollyhock** – manganese-deficient leaf.

9

6

7

6 Chinese lantern – interveinal chlorosis caused by manganese deficiency.

7 Chinese lantern – symptom confined to the youngest leaves, in this instance those just behind the floral buds.

8 *Nandina domestica* – pale green chlorosis of younger leaves with veins highlighted in green.

9 *Viburnum* – interveinal chlorosis caused by manganese deficiency is most developed in the youngest leaves.

8

10

12

11

10 ***Ficus benjamina*** – manganese deficiency symptoms appear on young and recently mature leaves.

11 **Waratah** – light brown-coloured necrotic lesions develop between the veins of recently mature leaves which remain dark green.

12 ***Banksia robur*** – the youngest leaves develop a light green interveinal chlorosis (compare with healthy leaf on right). Major veins are highlighted in green producing an iron deficiency-like pattern.

Zinc deficiencies are rare in ornamental crops in New South Wales where soils in the main growing areas tend to be acidic and well supplied with zinc. Problems with container plants have also been minor probably because zinc is routinely added during preparation of media and because zinc additions from other sources, including fungicidal sprays and irrigation water, are usually adequate for growth. Over-liming and use of high rates of phosphorus and nitrogen fertilisers may induce zinc deficiency especially if zinc supply is marginal.

Function Zinc has an important role in the formation and activity of chlorophyll and in the functioning of several enzymes and the growth hormone auxin. The severe stunting of leaves and shoots in zinc-deficient crops is a consequence of low auxin levels in tissue. The plant requirement for zinc is in the medium range for micronutrients, and around 30 ppm is usually satisfactory in leaves.

Symptoms Deficient plants show an overall paleness and a distinctive interveinal yellowing of the youngest leaves. Leaves are small and distorted. In some crops, only half of the leaf or leaflet develops fully, producing a 'sickle' or 'C' shape. Plants are stunted and prone to multiple branching. Shoot length is reduced and leaves are clustered near the growing tip (called 'rosetting'). Leaf symptoms are expressed earlier and strongest when the weather is bright and sunny and may not be very clear under the low light conditions found in shadehouses or igloos.

Corrective treatments A preventative soil application of a zinc compound such as zinc sulphate heptahydrate (23% Zn) or zinc oxide (60–80% Zn) broadcast and worked well into the soil prior to sowing is better than corrective sprays applied to the crop after symptoms develop. Apply 5–10 kg Zn/ha (20–40 kg zinc sulphate or 6–15 kg zinc oxide) depending on the severity of the deficiency, soil type and susceptibility of the crop. Soil treatments can remain effective for up to five years. For plants in soilless media, a root drench of zinc sulphate at 2.5 g/m^2 or 1.2 g/10 L is recommended. If zinc deficiency is identified in a developing crop, a foliar spray of 15–30 g zinc sulphate (23% Zn) in 100 L/ha as soon as possible after the first indications of deficiency in the crop, may reduce damage. Repeat sprays will be needed every two weeks if the crop is making active vegetative growth. Yield loss cannot be fully prevented due to the initial setback at the onset of deficiency. Zinc sulphate incorporated into soilless media at 70 g/m^3 should prevent deficiencies in container-grown crops. The Australian Standard for potting mixes requires 0.3–10 ppm of zinc in a DTPA extract. Crops which are regularly sprayed with zinc-containing fungicides, such as Zineb, Ziram or Mancozeb, rarely develop this deficiency.

1

2

3

1 **Rose** – leaves of zinc-deficient plant are small and distorted (healthy leaf on right).

2 **Gerbera** – strong interveinal chlorosis of the youngest leaves. Plant is stunted and flowering reduced.

3 **Waratah** – younger leaves develop pale green tips and are small and distorted.

4

5

4 **Waratah** – compared with healthy leaf (far left), younger leaves of zinc-deficient plants are small and distorted. The region near the leaf tip becomes chlorotic which contrasts strongly with the dark green colour of the remainder of the blade.

5 ***Banksia robur*** – interveinal chlorosis and distortion of the youngest leaves caused by zinc deficiency. Healthy leaf on left.

COPPER

Copper deficiency is not regarded as an important disorder of ornamental crops in Australia even though many soils and most organic media are deficient or low in copper. This is perhaps because copper sprays used to control fungal disorders and contaminants in fertilisers and irrigation waters supply sufficient copper for most plants. Copper is also routinely added during the preparation of potting media used in nursery production.

Copper deficiency may sometimes go unrecognised in ornamental crops. Yield losses and symptoms caused by a mild copper deficiency can be difficult to see unless healthy plants are available for comparison. Plant testing is not a sensitive guide to deficiency unless very young leaves are sampled. Being relatively immobile, copper accumulates in plant tissues with age. Copper concentrations in young leaves are better correlated with the current supply status than the concentration in mature leaves. Testing of soil or potting media can be helpful, however, the only sure way of identifying deficiency is to treat the crop with copper, leaving a few untreated pots as a control.

Copper is readily complexed by organic matter in soils and potting media and these forms have low water solubility. This limits the availability of copper to plants and reduces its movement in the growing medium. Copper deficiency is more likely to affect crops in peaty soils, or when large amounts of organic matter have been applied to a sandy or calcareous soil that is low in copper. Liming also increases the risk of a deficiency.

Copper deficiency is easily corrected by soil or foliar applications of copper. Because plants remove so little copper from the growing medium and because copper complexes are not easily leached out of the root zone, soil applications have a long residual effect. The low mobility of copper in plant tissues means that foliar sprays must be repeated at intervals to ensure that newly developed tissues are adequately supplied.

Function Copper is essential for photosynthesis, for the functioning of several enzymes, in flower and seed formation, and for the production of lignum which gives physical strength to shoots and stems as well as to vascular elements required for water movement in plants. Copper concentration in leaves of 5 ppm is satisfactory for most plants.

Symptoms The symptoms of copper deficiency can be quite variable between plant species, but symptoms that are common to many species include interveinal mottling, and tip burn and distortion of the youngest leaves, which results in rolling, bending or crinkling of mature leaves. Deficient plants are prone to wilting on warm days. Woody plants can develop a characteristic twisting or S-shaped growth of vegetative shoots which may or may not be accompanied by cracking and corky markings on the bark. Severe deficiency may cause stunting, delay flowering, reduce flower numbers and weights and may ultimately cause plant death. Copper-deficient field-grown crops are usually patchy, stunted and low yielding.

Copper deficiency can be difficult to diagnose from the symptoms which can be similar to the effects of frost and moisture stress. Also, yield and quality can be reduced without clear symptoms being produced. Leaf and soil analysis may be required to confirm a deficiency. DTPA extractable copper in the range 0.4–15.0 mg/L is desirable for organic potting media.

Corrective treatments Applications of 5–50 kg/ha of copper sulphate (bluestone), depending on species and soil type, well incorporated into the soil, can last for up to 10 years. As excess copper can be phytotoxic, use the lowest rate for sandy soils low in organic matter but the heavier peaty or marly soils may need 20–50 kg/ha. Copper deficiency in container-grown plants can be prevented by amending soilless media with 10–20 mg Cu/L. A foliar spray of 2 kg/ha of copper oxychloride, at either high or low volume, applied at an early growth stage, is a safe and effective correction for a growing crop. This treatment should be repeated two or three times in periods of active growth. Bordeaux spray can also be used, but unneutralised copper sulphate spray can cause leaf burn. Palms are particularly susceptible to this injury.

1

1 **Chrysanthemum** – strong interveinal chlorosis of the upper leaves. (Photo: Kevin Handreck.)

2

3

4

2 ***Abelia grandiflora*** – typical S-shaped growth of a shoot caused by copper deficiency.

3 ***Abelia grandiflora*** – photo showing corky stem of a copper deficient plant.

4 ***Erica gracilis*** – twisting and yellow chlorosis of the shoots on the left of the photo are typical of copper-deficiency. (Photo: Dr D. Alt.)

5

6

5 ***Picea amorika*** – copper-deficient plant (left) is pale and has an abnormal form with shoots growing down. (Photo: Dr D. Alt.)

6 **Silverbeet** – copper-deficient leaves have developed a strong yellow chlorosis with the veins highlighted in green. Growth of younger leaves is distorted. (Photo: Kevin Handreck.)

Elements harmful in excess (toxicities)

Any element taken up in excess of plant requirement can cause injury, however, toxicities of sodium, chloride, nitrogen, phosphorus, manganese, copper, zinc, boron and fluorine are most common in cultivated plants.

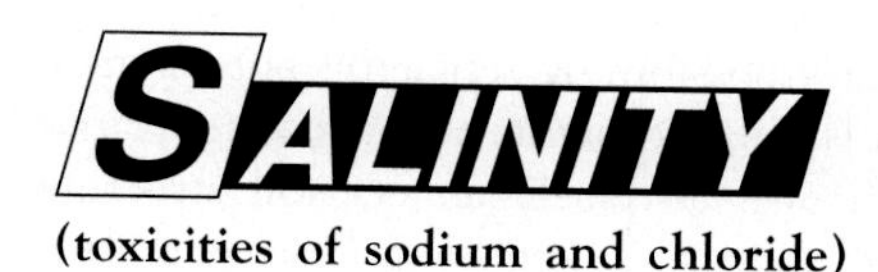

(toxicities of sodium and chloride)

High concentrations of salt in the soil solution are damaging to plants because they reduce water availability to roots (osmotic stress), and because absorbed salts can impair cell processes (specific ion toxicity). Salt may be absorbed through the leaves as well as by the roots so toxicities can occur even when soils are not particularly saline.

Salts of sodium (Na^{+}), chloride (Cl^{-}), calcium (Ca^{2+}), magnesium (Mg^{2+}), iron ($Fe^{2,3+}$), boron (B^{-}), sulphur (SO_4^{2-}) and carbonate (HCO_3^{-}) are usually present in saline waters and soils. Of these, Cl^{+}, B, and Na^{+} are the most toxic and do the most damage to crops. Soluble fertilisers also add to the total salt levels in the soil or growing medium, increasing the osmotic stress on plants and occasionally causing toxicity. (See following section on fertiliser toxicity.)

The concentration and the specific toxicity of individual salts in the soil solution or irrigation water are the main factors which determine just how much damage is caused by salinity. A plant's capacity to endure saline conditions is influenced by its nutrient status, the physical conditions and distribution of salts in the growing medium, method of irrigation, weather conditions, current growth rate, crop maturity, variety and rootstock. Often hot weather, drying winds or a temporary water shortage is needed before a crop expresses clear symptoms of salt stress.

Salt-tolerant plants (Appendix 2) have the ability to minimise ion accumulation in sensitive tissues. They do this by excluding toxic ions at the root surface, or by depositing excess minerals in older leaves or in stems. However, even those plants which have an effective root barrier to salts can be injured if significant quantities of Na^{+} or Cl^{-} are absorbed by leaves. The main sources of foliar-absorbed salts are sea spray and saline irrigation water applied through sprinklers. Spray irrigated crops are at greatest risk of salinity damage in warm, drying weather or when short, frequent irrigations in the heat of the day, lead to a build up of salts on the leaf surface. Foliar uptake of salts from irrigation water may be reduced by using under-canopy sprinklers which avoid leaf wetting, by irrigating at night and by giving longer, less frequent irrigations to minimise salt deposition on leaves by evaporation.

SODIUM AND CHLORIDE

Although sodium is not an essential element it can benefit the growth of some plant species possibly by replacing potassium in osmotic regulation of cells. Sodium is toxic to plants at low concentration in tissue. High sodium in the soil also adversely affects soil structure, particularly in clayey soils. Although salinity is a problem in large areas of Australia, sodium toxicities are rare in ornamental crops because they are normally grown where good quality water is available for irrigation. In general, woody plants are more sensitive to sodium toxicity than succulents. For most ornamental plants, the symptoms of sodium toxicity cannot be distinguished easily from those of chloride toxicity or moisture stress. The concentration of sodium in healthy leaves is normally less than 0.1% and toxicity is usually indicated by levels above 0.5%.

Chlorine is an essential element but it is only required in trace amounts. It is far less toxic than sodium or even iron and is the only trace element which can be accumulated to percentage levels without harming the tissues of most plants. Chloride toxicity is more common in crops than sodium toxicity because it is less readily excluded by roots. Once absorbed, chloride moves in the transpiration stream to the plant tops where it accumulates preferentially in mature and old leaves. Most plants can tolerate up to 1% chlorine in leaves before expressing toxicity symptoms.

Symptoms Plants affected by salinity have a reduced growth rate and are inclined to wilt in warm weather. They have dull bluish green foliage and often thicker more succulent leaves. Severe salt stress is indicated by tip burn and/or yellowing or scorching of the margins of mature leaves which is followed by leaf fall and shoot dieback. Tip burn is usually more prevalent in hot, drying weather and when high rates of fertiliser have been used.

While scorching on older leaves is caused by high chloride, damage to the young leaves and growing points is more often due to a shortage of water or calcium. Symptoms of salt injury, drought, potassium deficiency and fertiliser burn are easily confused, so leaf and soil analysis may be needed to confirm a diagnosis.

Corrective treatments With field-grown crops, irrigation can be used to leach salt from the root zone but adequate drainage must be provided to prevent the watertable rising.

Amending heavy-textured soils with gypsum can assist the leaching of salt from the root zone. Good quality water should be used for irrigation and so new sources of water should always be tested for salinity. Water quality guidelines are based on measurements of electrical conductivity (EC) as an indicator of osmotic stress (Table 8) and on the concentrations of individual salts which cause toxicity (Table 9).

Salts accumulate in container-grown crops when the supply from irrigation water and fertilisers exceeds losses from the medium through plant uptake and leaching. Evaporation tends to concentrate salts near the surface of the medium, particularly when subirrigation is used. The risk of

Table 8 General criteria for the salinity of irrigation waters

Level	Electrical conductivity (dS/m)	Total soluble salts (mg/L)
1	0–0.28	0–175
2	0.28–0.8	175–500
3	0.8–2.3	500–1500
4	2.3–5.5	1500–3500
5	Above 5.5	Above 3500

Source: Australian Water Resources Council (1974)

Irrigation

Level 1 Low salinity water can be used with most crops on most soils, with all methods of water application and with little likelihood that a salinity problem will develop. Some leaching is required but this occurs under normal irrigation practices except in soils of extremely low permeability.

Level 2 Medium salinity water can be used if a moderate amount of leaching occurs. Plants with medium salt tolerance can be grown, usually without special practices for salinity control. Sprinkler irrigation with the more saline waters in this group may cause leaf scorch on salt-sensitive crops, especially at high temperatures in the day time and with low water application rates.

Level 3 High salinity water cannot be used on soils with restricted drainage. Even with adequate drainage, special management for salinity control may be required, and the salt tolerance of the plants to be irrigated must be considered.

Level 4 Very high salinity water is not suitable for irrigation under ordinary conditions. For use, soils must be permeable, drainage adequate, water must be applied in excess to provide considerable leaching, and salt-tolerant crops should be selected.

Level 5 Extremely high salinity water may be used only on permeable, well drained soils using good management, especially in relation to leaching and for salt-tolerant crops, or for occasional emergency use.

Table 9 Guidelines for use of irrigation water containing sodium, chloride or boron

Salt	No hazard	Restricted use	Unsuitable[3]
Root uptake[1]			
Sodium[2] (millieq/L)	<2	2–18	>18
Chloride (mg/L)	<178	178–355	>355
Boron (mg/L)	<0.5	0.5–2	>2
Foliar uptake[4]			
Sodium (mg/L)	<70	>70	–
Chloride (mg/L)	<100	>100	–

[1] Data sourced from State Pollution Control Commission discussion paper on Water Quality Criteria for New South Wales, November 1990.

[2] Sodium adsorption ratio (SAR) = $\frac{[Na]}{[0.5\ ([Ca^{2+}] + [Mg^{2+}])]^{0.5}}$

[3] Water only suitable for tolerant crops.

[4] Based on data from Division of Agricultural Sciences, University of California, 1979.

salinity problems in container crops can be reduced by using the best possible quality water for irrigation, matching fertiliser rates to plant requirement and irrigating slightly in excess of requirement to leach the unused salts from the pot. A leaching fraction of around 20% is adequate. Saline potting mixes should never be allowed to become too dry as this will concentrate salts in the soil solution and make the problem worse. Mulches, potting media or potting media components should be tested for salinity and leached if necessary before use. Media derived from mushroom compost, seaweed and some sources of peat and coconut fibre need to be examined for residual salinity.

Soluble salt levels in irrigation water and in soil and plant media can be monitored to give an early warning of salinity. The Australian Standard for potting mixes specifies that the electrical conductivity of a 1:1.5 v/v water extract should not exceed 1.0 dS/m for seedling mixes and 1.8 dS/m for other mixes. DTPA extracts from seedling and orchid mixes should contain less than 60 ppm of sodium or chloride. Up to 100 ppm of sodium or chloride is permitted in other mixes. The electrical conductivity of a water extract (1:1.5 v/v) obtained from soilless media can be used to predict the salinity hazard to plants (Table 10).

Table 10 Potting media salinity guidelines.[1] Values based on the electrical conductivity (EC) of a 1:1.5 v/v water extract

Salinity hazard	EC (dS/m)
Nil	<0.7
Low	0.7–1.2
Moderate	1.3–1.8
Fairly high	1.9–2.7
High	2.8–3.6
Very high	>3.6

[1] Arnold Bik, R. and Boertje, G.A. (1975). *Acta Horticulturae* **50**, 153.

1

2

5

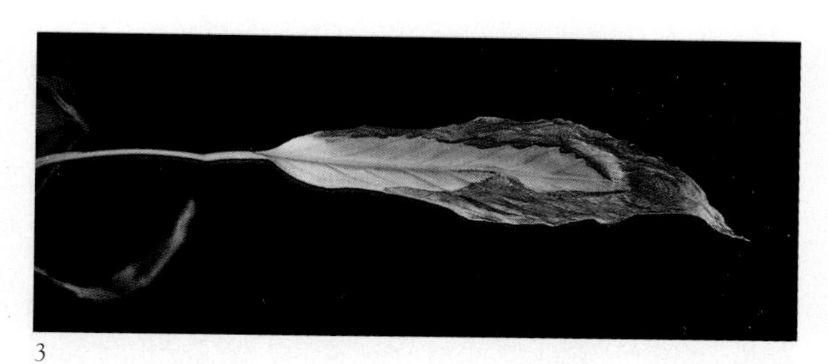

3

4

1 **Gloxinia** – plants grown under saline conditions produce dark green foliage and, in time, develop a brown scorch on mature leaves. Scorching commences as spots between the veins and is most severe near the leaf tip.

2 **Hollyhock** – mature leaves initially darken and then develop light green patches between the veins. A narrow band of necrotic tissue is sometimes present on the edges of the blade.

3 **Spathiphyllum** – burning of a mature leaf caused by chloride toxicity.

4 **Maple** – light brown burn commencing at the leaf tips due to excess chloride. The symptom spreads towards the petiole and is normally preceded by an interveinal chlorosis.

5 **Hibiscus** – marginal scorch and premature yellowing of mature leaves caused by salt toxicity.

6

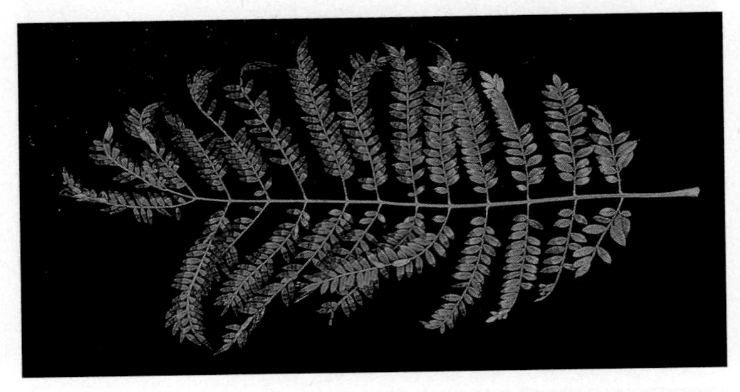

7

6 Ornamental fig – With chloride toxicity, large areas of tissue on mature leaves become necrotic and dry off to a grey–brown colour. These scorched areas are often irregular in shape but generally start near the edges of a leaf. The transition between dead and healthy tissue is defined by a dark band of water-soaked tissue.

7 Jacaranda – a brown scorch on leaflets is the primary symptom of chloride toxicity. The injury commences at the tips of the leaflet and is preceded by a subtle lightening in the colour of the affected tissue.

8

9

8 ***Grevillea*** **cv. Silky oak** – scorching and death of older leaves caused by use of irrigation water high in chloride.

9 Waratah – mature leaf symptom following treatment with sodium chloride. Leaf colour darkens and then clears from the margins in response to the local accumulation of chloride. Chlorotic areas then burn.

10 ***Howea forsteriana,*** **Kentia palm** – scorch symptom produced when plants were irrigated with water containing sodium chloride. Note the lighter area of tissue in the transitional zone between healthy and affected areas of the blade.

10

MANGANESE

Manganese toxicity in soil-grown crops is usually associated with acidic growing conditions or waterlogging. Where soilless media are used, toxicity can also be caused when steam sterilisation or heating of media releases biological forms of manganese, by the use of manganese containing materials such as fly-ash or rice hulls, or the over-use of trace elements in the nutrient program. Manganese toxicity occurs on a wide range of soil types. In New South Wales, the red basaltic soils of the north coast and southern tablelands have high available manganese levels but toxicities are also found in many other localities on a range of soils, including some sandy soils. Waterlogging or heating of bare summer-fallowed soils can increase manganese solubility in soil. Seedlings, especially stocks and petunia, are most susceptible to manganese toxicity particularly during periods of slow growth.

Symptoms Manganese toxicity causes a brown burn on the oldest leaves, commencing either at the margins or at the tip of the leaf blade. Initially, discrete dark necrotic spots appear near the veins of older leaves and occasionally on the petioles and stems. These spots are associated with a localised deposit of manganese dioxide at the end of a vein. As the spots coalesce, larger areas of necrosis are formed, and leaves higher on the shoot also begin to show symptoms. Older leaves may eventually turn yellow. Many crops, including chrysanthemum, roses, carnations and gerbera, also develop an iron-like chlorosis of younger leaves. With petunias and stocks both young and old leaves become chlorotic around the margins. Manganese toxicity can reduce vegetative growth and flower production.

Corrective treatments Manganese toxicity can be alleviated by correcting soil acidity and improving drainage. With field-grown crops, agricultural lime can be applied at the rate of 1 tonne per hectare or more, depending on the pH and buffering capacity of the soil. As a general rule, heavier textured soils need more lime than sandy soils. Liming rates for adjusting the pH of soilless media vary according to the nature and proportions of the components in the mix. In peat-based mixes, 2–3 kg/m^3 of a high grade lime or 5 kg/m^3 of dolomite is normally adequate to lift the pH to about 6.0–6.5 (1:1.5 water), however, a small trial should be carried out to establish the precise amount of lime required in a mix. Nutrient solutions or irrigation water containing more than 28 ppm manganese may cause toxicity in sensitive crops.

1

2

3

1 Rose – yellow–green chlorosis of the newer growth following treatment with excess manganese.

2 Gerbera – manganese toxicity induces a strong yellow–green chlorosis of younger leaves. Veins remain dark green until the condition is well advanced.

3 Hollyhock – iron-like chlorosis of the youngest leaves caused by manganese toxicity.

4

5

6

7

4 ***Howea forsteriana,* Kentia palm** – iron deficiency pattern on the youngest frond was induced by manganese toxicity.

5 **Waratah** – young, middle and old leaves from a plant receiving excess manganese showing the major symptoms of this disorder – uniform yellow–green chlorosis of the youngest leaves and yellowing and burning of the oldest leaves.

6 ***Banksia robur*** – strong iron deficiency patterns in the youngest leaves is the dominant symptom of manganese toxicity in this species.

7 ***Grevillea* cv. Royal Mantle** – reddish brown spots best seen on the undersurface of older leaves are a symptom of manganese toxicity in this species.

COPPER

Copper is potentially more toxic to plants than either manganese or zinc but field occurrences of toxicity are rare. This is because most horticultural soils and growing media contain very little copper and because the availability of any added copper is reduced by fixation by organic matter.

Copper toxicity has been reported in soil-grown crops from areas where Bordeaux sprays have been used over many years, such as old citrus orchards and vineyards. At these sites, copper tends to accumulate in the top few centimetres of soil. Soil from mine sites and silt dredged from contaminated waterways can contain high concentrations of copper capable of causing toxicity in reclamation plantings. Copper is more available to plants at low pH and so toxicities are more likely where management practices acidify the soil.

In soilless growing systems, plants can obtain excess copper from irrigation water, from farm chemicals, and from the growing medium. Water or nutrient solutions may pick up copper from pipes or brass fittings. This source is more likely to cause toxicity when the growing medium is inert, such as rockwool, perlite or sand, or where there is no medium as in solution culture (NFT). The quantities of copper supplied in the nutrient program are normally too small to cause toxicity, except where high rates are used in error. A misplaced decimal point can lead to a 10 or 100 fold increase in rate. Fungicidal sprays are a more common source of excess copper and can cause toxicity in sensitive plants like aphelandras, scheffleras, and most palms and ferns. Finally, the growing medium may be contaminated with copper. High concentrations of copper are sometimes present in sewage sludges, municipal solid wastes, mulches and composts made from timber treated for pest control. These materials should be tested before use in a production system.

Copper is not readily translocated to plant tops so leaf analysis cannot give an early warning of developing toxicity. Copper is generally accumulated in the roots, and these become stunted. Copper is used in root pruning agents and, at the appropriate rate, has minimal effect on plant tops. However, when copper is available at high levels throughout the growing medium it suppresses root uptake of phosphorus and some trace elements as well as causing direct toxic effects on the plant. Iron deficiency is the most common disorder induced by copper toxicity.

Symptoms Copper has limited mobility in plant tissues, so symptoms of toxicity tend to show first at the site of uptake. The initial response to root uptake is stunting and thickening of new roots. Symptoms will eventually appear in the tops but these are primarily a response to root injury. Copper toxicity will slow the growth rate and cause stunting. New leaves will usually develop a pale green iron chlorosis and, if the toxicity is severe, the oldest leaves may burn from the margins and between the veins. When toxicity has been caused by foliar uptake, the injury symptom is localised and restricted to the tissue which the copper spray contacted.

Typically, this means that only one side of a plant or a part of a leaf blade is damaged. Spray injury is characterised by a brown burn in a droplet pattern which is apparent within a few days of an application. If the plant recovers, new growth is healthy and without symptoms.

Control Prevention is the best means of controlling copper toxicity, so all media or water which is of doubtful or unknown quality should be tested for copper before it is used. Appropriate rates of copper should be used in the fertiliser program (see section on copper deficiency).

Foliar sprays of soluble copper salts, such as $CuSO_4$, are likely to cause injury even at low concentrations. The risk of this happening is reduced by using copper oxychloride or a Bordeaux mixture.

Where copper toxicity has been identified, some reduction in plant losses may be achieved by liming or applying high rates of superphosphate. Incorporation of organic amendments into light, sandy soil may also limit damage.

Although it is not unusual for leaf analysis to reveal high zinc concentrations, toxicities are rare in ornamental plants. This is because much of the zinc comes from fungicidal sprays and is present as a relatively harmless surface deposit on leaves.

Most agricultural soils do not supply enough zinc to cause toxicity. Plants can obtain excess zinc if grown in sewage sludges, municipal wastes, silts from contaminated waterways and wastes from zinc mines and smelters. Additionally, water or liquid fertilisers may pick up sufficient zinc from galvanised pipes, roofs, tanks and support structures to cause toxicity. Before the introduction of plastic pots in nursery production, plants grown in galvanised metal trays or tubs would occasionally experience toxicity.

Plant uptake of zinc is favoured in acid conditions and in light-textured, low organic matter soils. Zinc is not very mobile in plants and tends to accumulate in roots. Plants suffering from zinc toxicity are prone to wilting because of root injury.

Uptake of some heavy metals and trace elements is reduced by high zinc concentrations in the growing medium. This antagonism is responsible for the iron chlorosis symptom in new leaves which is characteristic of zinc toxicity.

Symptoms The most prominent symptom of excessive root uptake of zinc is iron chlorosis or yellowing of young leaves. This can be accompanied by burning of mature leaves and root damage. Growth is retarded and plants may wilt even when soil moisture appears adequate. Symptoms appear gradually in a crop and recovery is slow.

When the injury is caused by foliar uptake of zinc from a spray or from contaminated water, the major symptom is an irregular pattern of necrotic spots on leaves. This injury is confined to those tissues directly contacted with zinc and usually appears within a few days of an application. If the plant recovers, any new growth will be healthy and free of symptoms.

Corrective treatments The availability of zinc for plants in toxic conditions can be reduced by raising the pH of the growing medium with lime, by using high rates of phosphorus fertiliser, and by incorporating organic matter into soils. Zinc is not readily leached from soil so the consequences of contamination are long lasting.

1

1 Rose – youngest leaves take on a uniform yellow–pink colour which contrasts with the dark green of older leaves.

2

2 Gerbera – the blades of leaves produced by plants receiving an excessive supply of zinc (right) are dark green and mottled, and greatly reduced in size.

3

4

3 **Gloxinia** – in response to zinc toxicity, the plant at left has developed an iron-like chlorosis of all leaves. Dark brown necrotic areas are also apparent near the base of the leaf blade. (Photo: Dr D. Alt.)

4 **Stock** – zinc accumulates around the edges of mature leaves causing bone-coloured necrotic lesions. These increase in size and coalesce until large areas of tissue have died. (Photo: Kevin Handreck.)

BORON

Boron is toxic to plants even at relatively low levels of supply. Of all the nutrients, it has probably the narrowest range between the minimum required for optimum growth, and what is ultimately damaging to tissues. Toxicities in commercial crops are often caused by the careless or incorrect use of boron products to prevent or cure a deficiency. Use of contaminated material, such as growing media, sawdust, mulch, municipal waste, sewage sludge or irrigation water, particularly from bores, is another possible cause. Some plants are less sensitive to toxicity than others (Table 11), and this seems to be related to how well boron can be excluded at the roots rather than to how well it can be tolerated once absorbed.

Table 11 Relative tolerances of some ornamental plants to excess boron[1]

Very sensitive (threshold <0.5 g/m^3)	
Mahonia aquifolium	Oregon grape
Photinia × fraseri	Photinia
Xylosma congestum	Xylosma
Elaeagnus pungens	Thorny elaeagnus
Viburnum tinus	Laurustinus
Ligustrum japonicum	Wax-leaf privet
Feijoa sellowiana	Pineapple guava
Euonymus japonica	Spindle tree
Pittosporum tobira	Japanese pittosporum
Ilex cornuta	Chinese holly
Juniperus chinensis	Juniper
Lantana camara	Yellow sage
Ulmus americana	American elm
Sensitive (threshold 0.5–1.0 g/m^3)	
Zinnia elegans	Zinnia
Viola tricolor	Pansy
V. odorata	Violet
Delphinium sp.	Larkspur
Abelia × grandiflora	Glossy abelia
Rosmarinus officinalis	Rosemary
Platycladus orientalis	Oriental arborvitae
Pelargonium × hortorum	Geranium
Moderately sensitive (threshold 1.0–2.0 g/m^3)	
Gladiolus sp.	Gladiolus
Calendula officinalis	Marigold
Euphorbia pulcherrima	Poinsettia
Callistephus chinensis	China aster
Gardenia sp.	Gardenia
Podocarpus macrophyllus	Southern yew
Syzygium paniculatum	Brush cherry
Cordyline indivisa	Blue dracaena
Leucophyllus frutescens	Ceniza

Table 11 Relative tolerances of some ornamental plants to excess boron[1] (continued)

Moderately tolerant (threshold 2–4 g/m^3)	
Callistemon citrinus	Bottlebrush
Eschscholzia californica	California poppy
Buxus microphylla	Japanese boxwood
Nerium oleander	Oleander
Hibiscus rosa-sinensis	Chinese hibiscus
Lathyrus odoratus	Sweet pea
Dianthus caryophyllus	Carnation
Tolerant (threshold 6–8 g/m^3)	
Raphiolepis indica	Indian hawthorn
Carissa grandiflora	Natal plum
Oxalis bowiei	Oxalis

[1] Species listed in order of increasing tolerance based on appearance as well as growth reduction. From the review by E.V. Maas in *Agricultural Salinity Assessment and Management*, Ed. K.K. Tanji, American Society of Civil Engineers, 1990.

Symptoms Once absorbed by roots, boron moves in the transpiration stream to the tops where it accumulates preferentially in mature and old leaves which have high transpiration rates. Once assimilated in tissue, outward movement of boron is limited, so large concentration gradients can develop over relatively short distances within a single leaf blade. Highest concentrations of boron are found near the ends of major veins – the leaf tip in species with parallel venation – and the leaf margins in species with a radial pattern of venation, and this is where symptoms first develop.

Boron toxicity symptoms consist of burning and/or interveinal chlorosis commencing at the tip or the margins of the oldest leaves. When the burning is marginal, the expanding leaves may become cupped. In chrysanthemums, the tissue between the veins turns yellow before becoming necrotic but, in roses, the veins often clear first. Some plants develop small, brown or even whitish necrotic spots irregularly spaced over the blade but always close to the veins. Dark spots may be surrounded by a chlorotic halo. If the disorder is persistent, the older leaves are shed and the symptoms gradually spread up the shoot until even the flower buds are damaged. Boron toxicity in palms causes a light tan necrosis, commencing at the tips of the oldest leaflets. In petunias, interveinal chlorosis is followed by a marginal burn of old and new leaves. All crops may suffer significant growth and yield reductions, including fewer buds, delayed flowering, shorter flower stems and smaller blooms before leaf symptoms become prominent.

Corrective treatments Boron is very soluble in water and can be leached from a well drained soil or potting medium to alleviate toxicity in a crop. However, avoidance is a better strategy, and every care should be taken to ensure that crops do not come in contact with potentially toxic sources of boron, which include treated wood shavings and sawdust or saline water. Manufacturers, recommendations should always be strictly followed when applying boron to crops. Concentrations of boron above

5 ppm in irrigation water or approaching 0.3 ppm in a 2 mM DTPA potting media extract (1:1.5 v/v), are toxic to sensitive crops like petunias. When topdressing borax or boric acid powder, all lumps should first be broken up to avoid pockets of high availability in the soil. After the application, the bed should be well watered. More even application can be achieved by dissolving the powder in water and applying it as a drench.

1

2

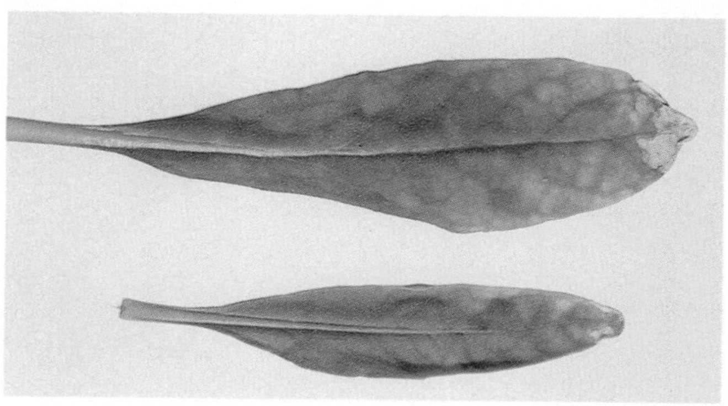

3

1 Rose – symptoms of boron toxicity are mainly found on older leaves and include vein clearing (a response to root death), necrosis and leaf fall. Necrotic lesions develop at the tip and around the margins of leaves and are generally associated with vein endings.

2 Gerbera – boron toxicity causes a pale green chlorosis of the youngest leaves and restricts leaf expansion. Mature leaves appear mottled with lighter areas between the veins. As the disorder worsens, patches of necrotic brown tissue appear on older leaves.

3 Stock – bleaching of leaves commencing from the tips is a symptom of boron toxicity.

4 **Viola** – stunting, yellowing of leaf margins and cupping of expanding leaves can be caused by excess boron. (Photo: Kevin Handreck.)

5 **Orchid** – tip burn and yellowing promoted by boron toxicity.

6 **Cape chestnut** – leaf scorch commencing from the tip of the blade and veinal clearing caused by boron toxicity.

4

5

6

NITROGEN (FERTILISER BURN)

Fertilisers (or manures) can burn plants if they are applied at too high a rate, placed too close to roots, or contact leaves directly during topdressing. Although high rates of any nutrient can be damaging, fertiliser injuries are mostly caused by soluble nitrogen salts. This can be confirmed by testing the burned leaf tissue or the tissue immediately adjacent to it for nitrate. Nitrate levels in healthy interveinal tissue are generally less than 1000 ppm whereas they may get as high as 10 000 ppm in injured tissue. High concentrations of ammonium or ammonia near the roots are also injurious to plants but this problem may not give rise to elevated leaf tissue levels. Seedlings planted into cold, waterlogged soils that are heavily fertilised with poultry manure, urea or ammonium fertiliser are most at risk of ammonium toxicity. Liming of newly fertilised ground encourages rapid ammonia release, which can lead to root injury.

Fertiliser burns to seedlings are often attributable to an uneven application of an appropriate amount of fertiliser. Azaleas and rhododendrons and most young transplants and seedlings in trays or boxes are sensitive to nitrogen fertilisers. Container plants moved indoors can suffer salt injuries if fertiliser rates are not reduced.

Symptoms Fertiliser injuries characteristically develop suddenly in a crop, generally within a few days of the application. Plant growth slows and the foliage becomes dark green and is prone to wilting. Young recently matured leaves are the first to show injury, developing translucent dark green areas within the blade which later become brown and dead. The leaves may droop and curl downward. Browning of vascular tissue in roots and stems with a narrow band of chlorosis and scorch on leaf margins are typical of ammonium toxicity. Plant roots become discoloured, turning from white to brown, and may eventually die and rot away. Other symptoms of fertiliser injury include leaf curl, defoliation, death and blackening of the shoot tips.

Corrective treatments Although fertiliser damage to tissues is often irreversible, plant recovery is helped if the growing medium is immediately leached with fresh water to remove salts from the root zone. The risk of ammonium toxicity in seedlings can be reduced by adding zeolite or other materials with strong cation exchange properties to the propagating medium. Fertiliser injury to subirrigated crops can be avoided by reducing fertiliser rates by about a third and by watering occasionally to leach excess nutrients from the pot. Fertiliser rates used on indoor plants under low light should be about a tenth of that required for intensive outdoor production.

1

2

3

4

5

6

1 **Rose** – wilting and leaf shedding caused by excess nitrogen fertiliser. Prior to abscission, leaves develop a reddish brown scorch. This symptom starts at the tip of the leaflet and progresses towards its base until the entire leaf blade is damaged.

2 **Gerbera** – burning of leaves due to an over-application of a nitrogen-based fertiliser.

3 **Chrysanthemum** – light brown scorch of the leaf margins starts at the ends of major veins. A spot test on damaged tissue will show high nitrate.

4 **African violet** – clearing and scorching of the margins of fully expanded leaves resulting from excess fertiliser.

5 **Bouvardia** – light brown necrosis of leaf margins due to nitrate toxicity. A narrow band of dark, water-soaked tissue defines the border between healthy and scorched tissue.

6 **Fiddler leaf** – fertiliser burn commences near the tip and spreads towards the base of the leaf. The initial symptoms are necrotic spots between the veins. These increase in size until most of the interveinal tissue is damaged.

7 Ficus – at high soil concentrations of nitrate, roots are burnt and will appear brown instead of white. The fine roots and root tips are most sensitive to injury. Damaged roots are more susceptible to pathogens.

8 Azalea – pinching and scorching of the upper leaves is a symptom of ammonium toxicity. Younger leaves may develop a mottled chlorosis similar to iron deficiency. (Photo: Dr D. Alt.)

9 Dracaena – scorching of the leaf tip due to excess nitrogen.

7

8

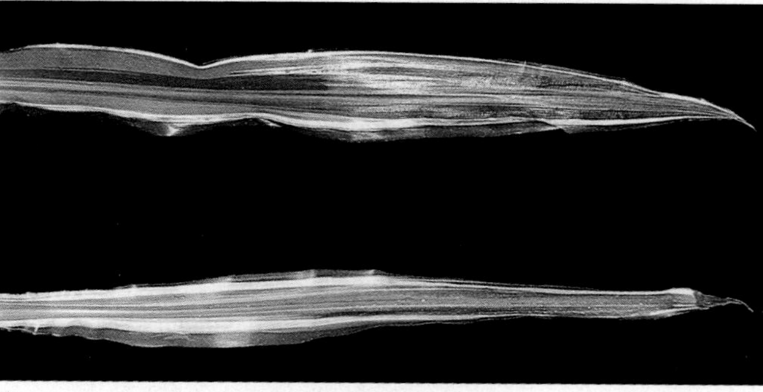

9

10

13

11

12

10 Palm – after an overdose of fertiliser, fronds yellow and then scorch from the tips of each leaflet.

11 Raphis palm – a yellow–green chlorosis of the leaf tip is followed by a dark, water-soaked lesion. After tissue death, the necrotic areas dry out.

12 Tree fern – light tan scorch on fronds caused by excess fertiliser.

13 *Protea neriifolia* – reddish brown scorch of the leaf tips due to excess nitrate.

Phosphorus

Toxicities develop when sensitive plants are exposed to high levels of soluble phosphorus as when base rates of fertiliser (normally superphosphate) are too high or when coated fertilisers release nutrients too rapidly. Plants in soilless media are more at risk from this toxicity than those in soil where chemical fixation reduces short-term peaks of available phosphorus. Phosphorus tends to be more toxic under alkaline growing conditions, perhaps because there is less iron for plant uptake or precipitation with phosphorus. Some genera within the Proteaceae, Fabaceae and Ericaceae and a surprising number of exotic species (Table 12) are sensitive to phosphorus toxicity and it is interesting to note that many of these plants are also susceptible to iron deficiency. Supplementing the growing medium with iron can reduce the severity of symptoms in sensitive plants.

Table 12 Plants sensitive to phosphorus toxicity have been found in these genera

Exotic plants	Australian native plants	
Azalea	Acacia	Grevillea
Calluna	Baeckea	Hakea
Camellia	Banksia	Hypocalymma
Chamaecyparis	Bauera	Isopogon
Cytisus	Beaufortia	Jacksonia
Elaeagnus	Boronia	Lechenaultia
Erica	Bossiaea	Leucadendron
Hydrangea	Brachysema	Leucospermum
Magnolia	Chorizema	Macadamia
Rhododendron	Conospermum	Protea
Senecio	Daviesia	Pultenaea
Skimmia	Dryandra	Serruria
Viburnum	Eucalyptus	Stirlingia
	Eutaxia	Telopea

Symptoms The main symptoms of severe toxicity are reduced growth, strong iron chlorosis of new leaves, burning and defoliation of older leaves and sudden plant death. Burning normally occurs around the margins of older leaves but it can develop in patches anywhere on the blade and is rarely preceded by yellowing. Bronzing, early leaf drop and root injury are other important symptoms of the toxicity. Mild toxicity may cause only a generalised paleness of the foliage or a reduction in flower quality which can be difficult to see if there are no healthy plants for comparison.

Leaf analysis can be used to confirm a diagnosis of phosphorus toxicity based on symptoms. Few reliable standards are available for sensitive plants, however, in some proteaceous plants phosphorus concentrations above 0.3% in a recently mature leaf or in the range 0.53%–0.95% in older symptom leaves are associated with toxicity.

When phosphorus supply is high, tolerant plants protect the young developing tissues by restricting phosphorus uptake at the roots (exclusion), by diverting luxury supplies of phosphorus to older leaves or shoots and/or by incorporating absorbed phosphorus into less damaging organic forms within the cell.

Rates of superphosphate commonly recommended for soil are generally too high for soilless media which have a lower capacity to fix phosphorus. In these media, preplant application of around 1 kg/m^3 of single superphosphate is normally adequate for non-sensitive plants. Phosphorus toxicity may develop when supply from controlled release fertiliser or liquid feeds exceeds 10 g P/m^3/month or when the DTPA extract (Australian Standard Method) of a mix exceeds 2–3 ppm phosphorus. Addition of 1.5 kg/m^3 of iron sulphate or 215 g/m^3 Fe EDDHA to the growing medium will give some protection against phosphorus toxicity.

1

1 Kangaroo paw – purpling and death of the tips of mature leaves. This is a common symptom and is not a reliable guide to phosphorus toxicity.

2 **Waratah** – the first indication of phosphorus toxicity is a lime green 'iron' chlorosis of the youngest leaves (two leaves at left of photo). As the condition progresses the tip and margins of the oldest leaves lighten in colour and then develop a red–brown necrosis (leaf on far right).

3 **Waratah** – leaves from a plant at an advanced stage of phosphorus toxicity have yellowed and burned from the margins.

4 ***Banksia robur*** – iron chlorosis of young leaves and burning of mature leaves following an excessive application of superphosphate.

2

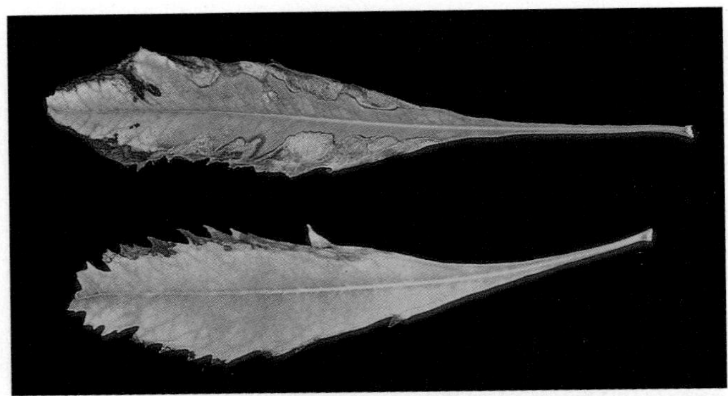

3

4

5

6

7

5 ***Banksia robur*** – oldest leaves are pale with an orange bleached region either side of the mid vein but not extending to the margins of the leaf. This tissue eventually dies and dries out.

6 ***Protea neriifolia* cv. Satin Mink** – younger leaves are chlorotic and older leaves develop dark brown necrotic lesions which spread until the entire blade is destroyed. Roots are also injured by phosphorus toxicity.

7 ***Protea neriifolia* cv. Satin Mink** – iron chlorosis caused by phosphorus toxicity is strongly developed in the plant on the left of the photo (healthy plant on right). Burning and shedding of older leaves and stunted growth are other important symptoms of this disorder.

8 Leschenaultia – plant on the left of the photo has developed a strong yellow chlorosis of shoot tips which is characteristic of phosphorus toxicity. (Healthy plant on right.)

9 Cyclamen – marginal burn on the edges of the blade caused by phosphorus toxicity.

10 Pittosporum – areas of necrosis develop on the margins of older leaves in response to excess phosphorus supply.

8

9

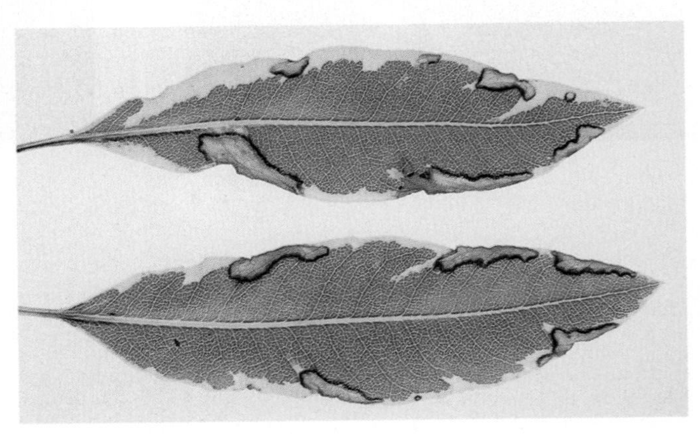

10

FLUORIDE

Fluorine is not essential for plant growth and causes injury to leaves and flowers at relatively low levels of supply. Plants may acquire fluorine from the air (atmospheric pollution), irrigation water, fertilisers (calcium nitrate, superphosphate), and liming materials. Plants in a number of genera are very sensitive to fluoride toxicity, including *Aspidistra*, *Calathea*, *Chlorophytum*, *Cordyline*, *Dracaena*, *Gladiolus*, *Lilium*, *Hybrids*, *Maranta*, *Spathiphylum*, *Stromanthe* and *Yucca*. Differences in plant susceptibility are due to differences in root uptake and tissue tolerance of absorbed fluoride (Appendix 5).

The symptoms of fluoride toxicity in most crops are similar whether uptake has been through the leaves or through the roots. The typical leaf symptom is a red–brown scorch commencing from the tip or margins of the blade and intruding between the veins towards the midrib in leaves with a reticulated pattern of veins. The boundary between healthy and dead tissue is usually sharply delineated often by a narrow water-soaked region.

Easter lilies grown in a fluoride-contaminated medium develop half-moon-shaped light tan to dark brown necrotic areas towards the tips of lower leaves. Half to two-thirds of the leaf blade may eventually become scorched. At high levels of supply bulb yield may decline. Hybrid lilies initially develop water-soaked margins near the apex of basal leaves. These areas later become scorched.

Flowers of susceptible crops including gerbera, gladiolus, freesia, rose, poinsettia and chrysanthemum that are kept in vase water containing as little as 1 ppm F may develop toxicity symptoms on petals. (Domestic water is commonly treated with 1–2 ppm F to improve dental health.) Gerberas develop a light brown scorch on the tips of petals which in time spreads towards the base. The extent of this injury, and the numbers of petals affected, increases with time as fluoride is accumulated by the transpiring tissues. Sensitive varieties may develop symptoms within the first 24 hours of the flowers being placed in the vase water.

Corrective treatments Damage from elevated atmospheric fluoride may be reduced by spraying plant foliage with lime water, and liming of contaminated soil or growing media to raise the pH above 6.5 will reduce root uptake of fluoride. Gypsum or activated carbon can be used to reduce the availability of fluoride to roots in plants which prefer acidic growing conditions. A small amount of soil added to a potting mix will also reduce fluoride uptake by plants. Superphosphate should not be used with very sensitive plant species as it can contain 1%–2% fluoride. Fluoride toxicity damage to cut flowers can be prevented by using rainwater, or deionised or distilled water in the vase solution.

1 **Gerbera** – toxic levels of fluoride absorbed from vase water produce a brown burn on petal tips.

2 **Gladioli** – fluoride toxicity caused by uptake from a vase solution is indicated by a horseshoe-shaped clear area on petals.

3 **Petunia** – fluoride absorbed from the air causes a bruise-like symptom around the edges of petals.

1

2

3

Some non-nutritional symptoms

Non-nutritional stresses can also cause nutrient-like symptoms in leaves, flowers or other tissues. However, the resemblance is often only superficial and basic differences in the patterns can be seen when the specimen is examined closely.

Non-nutritional symptoms can arise from environmental stresses like frost, heat, drought, waterlogging, or wind, or from physiological stress like root injury or spray burns, herbicides, genetic abnormalities, virus infection or insect injury.

Purpling Purpling of the stems and foliage is a stress response exhibited by many plants. It can be induced by cold weather, root injury, transplanting shock, vascular disease and viruses. Purpling is often a symptom of nitrogen and phosphorus deficiency but, by itself, it is not a reliable guide to these disorders.

Vein chlorosis The terms 'vein chlorosis', 'vein clearing' or 'yellow vein' are used to describe the loss of green colour from the midrib and major veins of leaves. The interveinal tissue remains green, though it may become slightly paler than normal. This pattern is contrary to most nutritional patterns where colour loss begins between the veins. Although vein clearing can be a symptom of nitrogen deficiency in some crops like roses, it more often indicates root injury, waterlogging, or damage from herbicides or disease.

Herbicides Herbicides taken up through the roots or leaves can cause a variety of symptoms, including stunting or distortion of leaves or shoots, vein clearing and chlorotic leaf patterns.

The chlorosis caused by many herbicide sprays is different to a typical nutrient deficiency symptom, in that pattern symmetry is usually absent and the chlorotic areas are unrelated to leaf venation. Systemic herbicides can produce a symmetrical pattern but these symptoms generally spread outwards from the veins, contrary to the development of most nutritional patterns. The colour of the chlorosis is often an unnaturally bright hue of yellow, orange, cream or white.

Root injury The presence of more than one nutritional symptom in leaves can be a sign that roots are damaged or diseased or that the vascular system has been impaired. Under these conditions, nutrient supply to the tops is restricted and very often the first symptoms seen is an iron chlorosis of the young leaves. Other symptoms are produced in the older leaves and, as might be expected, the more mobile nutrients like nitrogen and magnesium are the first to become limiting. Vein clearing can be a sign of root injury from chemicals or gas leaks.

Spray burn Sprays of trace elements and other nutrients or pesticides can injure leaves, flowers or fruit if they are applied incorrectly – at the wrong strength or time, during heatwave conditions, or to plants which are moisture-stressed. Injury can occur if chemicals are not correctly mixed or kept agitated in the vat during spraying, or where several incompatible compounds are combined in the one spray, and react to produce a phytotoxic residue. Softly grown glasshouse or shadehouse plants can be injured by spray concentrations normally safe for field-hardened plants.

Though spray injury can vary with the chemical used, the conditions at spraying, the equipment used and the type and condition of the plant and tissue maturity, this type of damage can be recognised from the injury pattern – often in the shape and distribution of the spray droplets. The sudden appearance of symptoms is a further indication. Burns usually show within days of the spray application with many plants developing the symptoms simultaneously. Nutrient disorders tend to show up to different degrees throughout the crop, the symptoms first appearing in patches, then gradually spreading.

Environmental stress Symptoms caused by adverse environmental conditions, such as frost, hail, heat or wind, are also usually visible within hours or, at most, a few days of some unusual weather.

Frost Extreme cold, particularly frosts, can injure plants causing death and collapse of tissue which then appears blackish in colour. The problem is greatest in periods of unseasonally warm winter weather when plants are still growing rapidly and producing soft new growth.

Leaves or other plant parts develop dry necrotic areas on the exposed parts. Young plants are most at risk with the growing point often being killed. In some plants, mature leaves protect the growing point and young tissues. Exposed leaves wilt and later develop dry burnt patches. The midribs and petioles of large succulent leaves can become brittle and develop splits or cracks. Frost damage is usually worse in the low lying parts of the field or growing area.

Drought and heat If the stress is only moderate and for short duration, temporary wilting may be all that happens, leaving no permanent symptom, but a severe moisture stress will cause irregular-shaped burning of the outer parts of leaves. These leaves may yellow and die. Sunscald, caused by sudden heatwave conditions, can produce similar effects on leaves even when soil moisture is adequate. The distribution of healthy and damaged tissue is often related to the presence or absence of shading from protecting leaves. Hot, drying winds can induce a severe water stress even when the soil or potting medium is quite wet. Under these conditions, leaves will develop a light green to brown marginal burn. Plants grown under shade and moved to the open can be injured on a hot day. Excessively high air temperature can develop in plastic or glass covered houses if ventilation is inadequate. Plants which have been heat stressed will often shed leaves several days later.

Water stresses (too little or too much) Symptoms of prolonged drought include wilting, stunting, yellowing and burning of older leaves, leaf drop and finally death of the whole plant. Symptoms of a short sharp water stress can be difficult to distinguish from the effects of high salinity, potassium deficiency and fluoride toxicity.

Plants that have experienced a moderate water stress on several occasions but have recovered without suffering tissue death produce dark blue–green coloured foliage which also appears dull and lifeless. More severe stresses cause necrosis or burning of leaf tips and margins. When the stress is sudden, tissue death occurs before senescence is possible and so burned areas of leaf have a grey–green colour and there is little immediate yellowing in associated living tissue.

Plants also produce symptoms in response to too much water. These are normally a consequence of root injury caused by low oxygen levels in the waterlogged soil or growing medium. Surprisingly the initial symptoms of excess water, wilting and leaf drop are the same as those for drought. In time, the veins of older leaves may clear while younger leaves can become chlorotic. Plant growth will slow and new tissue may be distorted. Affected plants are more susceptible to disease and will eventually die unless drainage is quickly improved.

Hail, raindrop, and water damage Hail can perforate or tear leaves or even defoliate a plant while large raindrops from a heavy squall will bruise very soft leaves. Cold water lying on leaves of succulent species like African violets can produce a characteristic pale chlorotic ring. Water droplets may also injure less sensitive plants if present on leaves during the heat of the day. Iron-stained irrigation water can leave a bronzy brown metallic sheen on leaves.

Windburn Wind effects include leaf tatter, distortion, abrasion from sand blasting and girdling (where the crown of the plant is rubbed against the soil). Windburn sometimes causes a silvery sheen on the wind-exposed surface of leaves.

Genetic abnormalities These tend to be found as a random scatter of affected plants throughout a crop rather than in patches as usually happens with nutritional disorders. Isolated plants may be stunted and distorted or show irregular yellow patterns in leaves, or abnormal shape or colour of flowers or fruit.

Air pollutants Ornamental plants are sensitive to a wide range of common air pollutants (Table 13) (Appendices 3–7).

Sulphur dioxide, produced when heating fuels with a high sulphur content are burned, causes interveinal bleaching and a reduction in vein size in the bracts of poinsettia and leaf drop in roses. Elevated ozone levels cause reddish brown stippling on the upper surface of azalea leaves.

Fluoride toxicity is generally indicated by marginal or tip necrosis of leaves. Gladioli are particularly sensitive and develop an ivory, tan or brown necrosis which begins at the apex of the leaf. A narrow band of water-soaked tissue generally separates the healthy and affected regions. Other bulb plants including tulips, iris and lilies are also sensitive to fluoride. Fluoride toxicity caused by plant uptake from soil or water is discussed earlier (see under toxicities).

Mercury vapour and fumes from creosote and tar are toxic to some plant species as well as humans and use of these compounds within enclosed growing areas should be avoided.

Ethylene from motor vehide exhaust or produced by incomplete combustion in heaters can stunt and distort the growth of crops under cover. Ripening fruit and vegetables, foliage and flowers and plant disease organisms produce ethylene which can dramatically reduce the vase life of

Table 13 Air pollutants, sources, and sensitive ornamental plants[1]

Air pollutant	Source	Some sensitive ornamental plants
Ozone	Photochemical reactions in the atmosphere, storm centres, natural occurrence in the upper atmosphere.	See Appendix 3.
Sulphur dioxide	Combustion of fuel, petroleum and natural gas industry, smelting and refining of ores.	See Appendix 4.
Fluoride	Aluminium industry, manufacture of phosphate fertiliser, steel manufacture, brick plants, refineries.	See Appendix 5.
Peroxyacetyl nitrate (PAN)	Photochemical reaction in the atmosphere.	Dahlia, petunia.
Oxides of nitrogen	Exhaust gases of motor vehicles, combustion of natural gas, fuel oil and coal, refining of petroleum, incineration of organic wastes.	See Appendix 6.
Particulates	Combustion of coal, petrol and fuel oil, cement mills, lime kilns, incinerators.	
Ethylene	Leaky natural gas heaters, natural occurrence, combustion of coal and oil, motor vehicles, refuse burning.	See Appendix 7.
Ammonia	Leaks or breakdowns in industrial operations, spillage of anhydrous ammonia.	Petunia, sunflower.
Chlorine	Refineries, glass industry, scrap burning, accidental spills.	Rose, tulip, zinnia, violet, iris.
Hydrogen chloride	Same as chlorine.	Chrysanthemum.

[1] Summarised by Strider from Krupa, S.V. and Pfleger, F.L. (1975). Effects of air pollutants on ornamental plants. *Florists' Rev.* **155(4025)**, 24–25, 42–45.

some cut flowers. Typical symptoms include petal wilt (carnations), shattering, flower abscission (snapdragons), wilting and bleaching of sepals (orchids), failure to flower, thick stems, short internodes, death of older leaves, and proliferation of vegetative shoots around the terminal flower bud (chrysanthemums).

Chlorine, used to control water-borne diseases and algae in irrigation water and recirculating hydroponic solutions, can injure plant roots. The main symptoms are a golden brown discoloration of the normally white roots but, in extreme cases, plant tops may wilt. Toxicity is more likely in systems where inert media such as rockwool, perlite or sand are used, or where there is no physical medium as in NFT, because chlorine is rapidly inactivated by organic matter. As a general rule, chlorine concentrations in the root zone should not exceed 2 ppm.

1

2

3

1 **Grevillea** – scorch symptom produced by strong sunlight combined with high air temperatures.

2 ***Howea forsteriana*, Kentia palm** – Plants grown under shade can be sunburnt when moved to full sun.

3 **Bromeliad** – burning and tissue breakdown caused by frost.

4

5

6

7

8

9

10

4 *Gingko biloba* – death of roots following an extended period of waterlogging leads to yellowing of older leaves and a reduction in the size of new leaves.

5 Alocasia – scorch symptom produced when water remains on leaves on a hot day.

6 Polyanthus – irregular unpigmented areas can develop on shaded leaves.

7 Blackbean – movement of healthy plants into low light conditions can lead to yellowing, vein clearing and leaf shedding.

8 Gerbera – stem splitting is a physiological disorder associated with fluctuating temperatures.

9 Gerbera – chemical spray burn to petals (flower on right).

10 White cedar – Vein clearing caused by uptake of a systemic herbicide.

11

13

12

11 Gerbera – clearing of interveinal tissue caused by root uptake of a systemic herbicide.

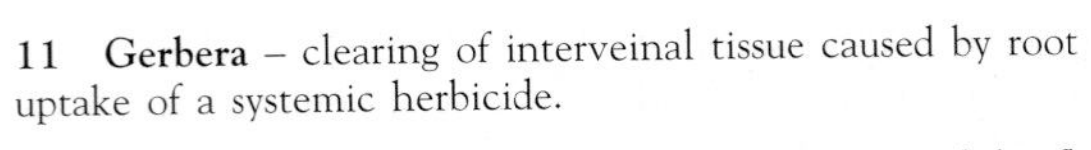

12 Lilium – hormone injury causing a distortion of the flower stem.

13 Carnation – powdered fertilisers or chemicals which lodge in the leaf axil following application can cause a localised burn.

14 Daphne – vein clearing and mottling of leaves in response to a viral infection.

14

15

17

15 Roses – viruses can produce spectacular and very regular symptoms in leaves because they can move within the vascular system.

16 Gerbera – broad mite injury to flowers is characterised by distortion of petals which appear cupped and marked. In a severe case, the bud may not produce a viable flower.

17 Gerbera – leaves from plants infested with broad mites are brittle and distorted and have a shiny reddish brown undersurface.

16

Useful references

1 Anon. (1993). Australian Standard for Potting Mixes AS 3743-1993. Standards Australia, Sydney.

2 Benton Jones, J., Wolf, B. and Mills, H.A. (1991). *Plant Analysis Handbook 1. Methods of Plant Analysis and Interpretation* Micro–Macro Publishing Inc, Athens, Georgia, USA. ISBN 1-878148-001.

3 Bergmann, W. (1992). *Nutritional Disorders of Plants: Development, Visual and Analytical Diagnosis*. Gustav Fischer Verlag, Jena, Stuttgart, New York. ISBN 1-56081-357-1.

4 Chase, A.R. and Broschat, T.K. (1991). *Diseases and Disorders of Ornamental Palms*. APS Press, Minnesota, USA. ISBN 0-89054-119-1.

5 Handreck, K.A. and Black, N.D. (1994). *Growing Media for Ornamental Plants and Turf*. University of New South Wales Press, Sydney, Australia. ISBN 0 86840 333 4.

6 Reuter, D.J. and Robinson, J.B. (1997). *Plant Analysis: An Interpretation Manual*. Second edition. CSIRO Publishing, Melbourne, Australia. ISBN 0 643 05938 5.

7 Winsor, G. and Adams, P. (1987). *Glasshouse Crops Vol. 3 Diagnosis of Mineral Disorders in Plants*. Chemical Publishing, New York. ISBN 0-8206-0322-8.

Appendices

Appendix 1 Leaf analysis standards

The leaf analysis standards that follow have been collected by the authors from a range of sources. The majority have come from scientific literature and from specialist publications such as books and trade journals, others are from unpublished research, and a third group are working values used by commercial laboratories. In some instances the standards for a single crop have been compiled from several of these sources.

Wherever possible the original source of the information has been acknowledged. See page 222 for the references quoted in the tables.

The presence of a standard in this publication does not mean that the information is reliable. Critical values should be treated as indicative and should be acted on cautiously until they have been proven by experience.

The superscript indicators [1] and * in the leaf analysis standard tables refer to the two footnotes below.

* Leaf analysis is not a reliable guide to iron deficiency because of surface contamination with dirt, or immobility of iron within the plant, or the presence of physiological inactive iron within tissues.

[1] Values for copper, zinc or manganese in leaves sprayed with fungicides or nutrient sprays containing trace elements cannot give a reliable guide to nutritional status even in washed leaves.

Abies alba
(Silver fir)

Sampling time One- or two-year-old needles
Plant part Uppermost lateral shoot

Ref: 2

Nutrient[1]	Deficient	Low	Normal	High	Excess
Nitrogen % (N)			1.30–1.80		
Phosphorus % (P)			0.13–0.35		
Potassium % (K)			0.50–1.10		
Calcium % (Ca)			0.40–1.20		
Magnesium % (Mg)			0.15–0.40		
Sulphur % (S)					
Sodium % (Na)					
Chloride % (Cl)					
Copper ppm (Cu)			5–10		
Zinc ppm (Zn)			15–60		
Manganese ppm (Mn)			50–500		
Iron* ppm (Fe)					
Boron ppm (B)			20–50		
Molybdenum ppm (Mo)			0.06–0.25		

Acer
(Maple tree)

Sampling time Current year's terminals at mid season
Plant part Most recently mature leaf

Ref: 2

Nutrient[1]	Deficient	Low	Normal	High	Excess
Nitrogen % (N)			1.70–2.20		
Phosphorus % (P)			0.15–0.25		
Potassium % (K)			1.00–1.50		
Calcium % (Ca)			0.30–1.50		
Magnesium % (Mg)			0.15–0.30		
Sulphur % (S)					
Sodium % (Na)					
Chloride % (Cl)					
Copper ppm (Cu)			6–12		
Zinc ppm (Zn)			15–50		
Manganese ppm (Mn)			30–100		
Iron* ppm (Fe)					
Boron ppm (B)			15–40		
Molybdenum ppm (Mo)			0.05–0.20		

Achillea millefolium
(Common yarrow)

Plant part Most recently mature leaf
Growth stage Flower buds visible

Ref: 1

Nutrient[1]	Deficient	Low	Normal	High	Excess
Nitrogen % (N)			2.66–3.08		
Phosphorus % (P)			0.34–0.43		
Potassium % (K)			5.04–5.56		
Calcium % (Ca)			0.75–0.96		
Magnesium % (Mg)			0.19		
Sulphur % (S)					
Sodium % (Na)					
Chloride % (Cl)					
Copper ppm (Cu)					
Zinc ppm (Zn)			37–48		
Manganese ppm (Mn)			60–68		
Iron* ppm (Fe)			127–147		
Boron ppm (B)			29–32		
Molybdenum ppm (Mo)					

Achillea
(cv. Coronation Gold)

Plant part Most recently mature leaf
Growth stage Flower buds visible

Ref: 1

Nutrient	Deficient	Low	Normal	High	Excess
Nitrogen % (N)			2.90		
Phosphorus % (P)			0.24		
Potassium % (K)			3.15		
Calcium % (Ca)			0.70		
Magnesium % (Mg)			0.18		
Sulphur % (S)					
Sodium % (Na)					
Chloride % (Cl)					
Copper ppm (Cu)					
Zinc ppm (Zn)			30		
Manganese ppm (Mn)			33		
Iron* ppm (Fe)			210		
Boron ppm (B)			35		
Molybdenum ppm (Mo)					

Acidanthera bicolor
(Abyssinian gladiolus)

Plant part Most recently mature leaf
Growth stage Flower buds visible
Ref: 1

Nutrient[1]	Deficient	Low	Normal	High	Excess
Nitrogen % (N)			2.50		
Phosphorus % (P)			0.29		
Potassium % (K)			1.54		
Calcium % (Ca)			0.47		
Magnesium % (Mg)			0.16		
Sulphur % (S)					
Sodium % (Na)					
Chloride % (Cl)					
Copper ppm (Cu)					
Zinc ppm (Zn)			50		
Manganese ppm (Mn)			32		
Iron* ppm (Fe)			110		
Boron ppm (B)			15		
Molybdenum ppm (Mo)					

Adenanthos sericeus

Plant part Most recently mature leaf
Ref: 17

Nutrient[1]	Deficient	Low	Normal	High	Excess
Nitrogen % (N)		<1.00	1.00–2.50		
Phosphorus % (P)		<0.06	0.06–0.28		
Potassium % (K)		<1.25	1.25–1.90		
Calcium % (Ca)			0.87		
Magnesium % (Mg)			0.19–0.25		
Sulphur % (S)			0.15–0.20		
Sodium % (Na)			0.25–0.54		
Chloride % (Cl)			0.25–0.36		
Copper ppm (Cu)			8–10		
Zinc ppm (Zn)			60		
Manganese ppm (Mn)			120–170		
Iron* ppm (Fe)			90–110		
Boron ppm (B)					
Molybdenum ppm (Mo)					

Adiantum pedatum
(Maidenhair fern)

Plant part Most recently mature leaf — Ref: 9

Nutrient[1]	Deficient	Low	Normal	High	Excess
Nitrogen % (N)		1.50–1.79	1.80–2.50	>2.50	
Phosphorus % (P)		0.18–0.29	0.30–0.60	>0.60	
Potassium % (K)		1.80–2.19	2.40–3.50	>3.50	
Calcium % (Ca)		0.08–0.11	0.12–0.30	>0.30	
Magnesium % (Mg)		0.15–0.24	0.25–0.40	>0.40	
Sulphur % (S)		0.15–0.19	0.20–0.40	>0.40	
Sodium % (Na)					
Chloride % (Cl)					
Copper ppm (Cu)		6–9	10–50	>50	
Zinc ppm (Zn)		16–24	25–200	>200	
Manganese ppm (Mn)		20–29	30–300	>300	
Iron* ppm (Fe)		30–49	50–300	>300	
Boron ppm (B)		16–24	25–50	>50	
Molybdenum ppm (Mo)					

Aechmea fasciata
(Silver vase plant)

Sampling time Non-flowering
Plant part Most recent fully developed leaf — Ref: 7, 19

Nutrient[1]	Deficient	Low	Normal	High	Excess
Nitrogen % (N)		1.20–1.40	1.5–2.0	>2.0	
Phosphorus % (P)		0.20–0.29	0.3–0.7	>0.7	
Potassium % (K)		1.20–1.40	1.5–3.0	>3.0	
Calcium % (Ca)		0.30–0.49	0.5–1.0	>1.0	
Magnesium % (Mg)		0.25–0.39	0.4–0.8	>0.8	
Sulphur % (S)		0.12–0.15	0.16–0.25	>0.25	
Sodium % (Na)					
Chloride % (Cl)					
Copper ppm (Cu)		4–6	7–30	>30	
Zinc ppm (Zn)		16–24	25–200	>200	
Manganese ppm (Mn)		30–49	50–300	>300	
Iron* ppm (Fe)		30–49	50–300	>300	
Boron ppm (B)	<15	16–24	25–50	>50	>76
Molybdenum ppm (Mo)					

Aeschynanthus pulcher
(Lipstick plant)

Plant part Most recently mature leaf — Ref: 9

Nutrient[1]	Deficient	Low	Normal	High	Excess
Nitrogen % (N)		1.70–2.09	2.10–2.8	>2.8	
Phosphorus % (P)		0.15–0.19	0.20–0.4	>0.4	
Potassium % (K)		2.00–2.49	2.50–3.3	>3.3	
Calcium % (Ca)		0.50–0.79	0.80–1.6	>1.6	
Magnesium % (Mg)		0.18–0.24	0.25–0.4	>0.4	
Sulphur % (S)		0.15–0.19	0.20–0.3	>0.3	
Sodium % (Na)					
Chloride % (Cl)					
Copper ppm (Cu)		6–9	10–50	>50	
Zinc ppm (Zn)		16–24	25–200	>200	
Manganese ppm (Mn)		30–49	50–300	>300	
Iron* ppm (Fe)		30–49	50–300	>300	
Boron ppm (B)		16–24	25–50	>50	
Molybdenum ppm (Mo)					

Aglaonema communtatum
(Chinese evergreen)

Plant part Most recently mature leaf — Ref: 7, 9

Nutrient[1]	Deficient	Low	Normal	High	Excess
Nitrogen % (N)		2.5–2.9	3.0–3.8	>3.8	
Phosphorus % (P)		<0.2	0.2–0.4	>0.4	
Potassium % (K)		2.5–2.9	3.0–4.5	>4.5	
Calcium % (Ca)		<1.0	1.0–2.0	>2.0	
Magnesium % (Mg)		0.22–0.29	0.3–0.6	>0.6	
Sulphur % (S)		0.13–0.19	0.2–0.4	>0.4	
Sodium % (Na)					
Chloride % (Cl)					
Copper ppm (Cu)		5–7	8–25	>25	
Zinc ppm (Zn)		17–19	20–200	>200	
Manganese ppm (Mn)		<50	50–300	>300	
Iron* ppm (Fe)		40–49	50–300	>300	
Boron ppm (B)	<15	15–24	25–75	>75	
Molybdenum ppm (Mo)					

Allamanda cathartica (Allamanda)

Plant part Most recently mature leaf — Ref: 7, 9

Nutrient[1]	Deficient	Low	Normal	High	Excess
Nitrogen % (N)		1.60–1.99	2.0–4.0	>4.0	
Phosphorus % (P)		0.15–0.24	0.25–1.0	>1.0	
Potassium % (K)		1.50–1.99	2.0–4.0	>4.0	
Calcium % (Ca)		0.50–0.75	0.76–1.5	>1.5	
Magnesium % (Mg)		0.20–0.24	0.25–1.0	>1.0	
Sulphur % (S)		0.13–0.19	0.20–0.4	>0.4	
Sodium % (Na)					
Chloride % (Cl)					
Copper ppm (Cu)		6–7	8–25	>25	
Zinc ppm (Zn)		15–19	20–200	>200	
Manganese ppm (Mn)		40–49	50–200	>200	
Iron* ppm (Fe)		40–49	50–200	>200	
Boron ppm (B)	<19	20–24	25–75	>75	
Molybdenum ppm (Mo)					

Alstroemeria

Plant part Most recently mature leaf — Ref: 9

Nutrient[1]	Deficient	Low	Normal	High	Excess
Nitrogen % (N)		3.20–3.69	3.70–5.60	>5.60	
Phosphorus % (P)		0.21–0.29	0.30–0.75	>0.75	
Potassium % (K)		3.00–3.69	3.70–4.80	>4.80	
Calcium % (Ca)		0.40–0.59	0.06–1.50	>1.50	
Magnesium % (Mg)		0.15–0.19	0.20–0.50	>0.50	
Sulphur % (S)		0.21–0.29	0.30–0.75	>0.75	
Sodium % (Na)					
Chloride % (Cl)					
Copper ppm (Cu)		2–3	4–50	>50	
Zinc ppm (Zn)					
Manganese ppm (Mn)		40–49	50–200	>200	
Iron* ppm (Fe)		100–149	150–300	>300	
Boron ppm (B)		10–12	13–50	>50	
Molybdenum ppm (Mo)					

Anthurium andraeanum
(Anthurium)

Plant part Most recently mature leaf minus petiole — Ref: 7, 9

Nutrient[1]	Deficient	Low	Normal	High	Excess
Nitrogen % (N)		1.02–1.59	1.60–3.0	>3.0	
Phosphorus % (P)		0.15–0.19	0.20–0.7	>0.7	
Potassium % (K)		0.70–0.99	1.00–3.5	>3.5	
Calcium % (Ca)		0.80–1.19	1.20–2.0	>2.0	
Magnesium % (Mg)		0.25–0.49	0.50–1.0	>1.0	
Sulphur % (S)		0.12–0.15	0.16–0.75	>0.75	
Sodium % (Na)					
Chloride % (Cl)					
Copper ppm (Cu)		<5	6–30	>30	
Zinc ppm (Zn)		<20	20–200	>200	
Manganese ppm (Mn)		<50	50–300	>300	
Iron* ppm (Fe)		<50	50–300	>300	
Boron ppm (B)	<15	16–24	25–75	>75	
Molybdenum ppm (Mo)					

Antirrhinum sp.
(Snapdragon)

Plant part Most recently mature leaf — Ref: 6

Nutrient[1]	Deficient	Low	Normal	High	Excess
Nitrogen % (N)			4.0–5.3		
Phosphorus % (P)			0.2–0.6		
Potassium % (K)			2.2–4.1		
Calcium % (Ca)			0.5–1.4		
Magnesium % (Mg)			0.5–1.0		
Sulphur % (S)					
Sodium % (Na)					
Chloride % (Cl)					
Copper ppm (Cu)			5–15		
Zinc ppm (Zn)			30–55		
Manganese ppm (Mn)			60–185		
Iron* ppm (Fe)			70–135		
Boron ppm (B)			15–40		
Molybdenum ppm (Mo)					

Aphelandra squarrosa
(Zebra plant)

Plant part Most recently mature leaf — Ref: 9

Nutrient[1]	Deficient	Low	Normal	High	Excess
Nitrogen % (N)		1.06–1.99	2.0–3.0	>3.0	
Phosphorus % (P)		0.15–0.19	0.2–0.4	>0.4	
Potassium % (K)		0.90–1.09	1.1–2.0	>2.0	
Calcium % (Ca)		0.3–0.59	0.6–2.0	>2.0	
Magnesium % (Mg)		0.3–0.49	0.5–1.0	>1.0	
Sulphur % (S)		0.15–0.19	0.2–0.3	>0.3	
Sodium % (Na)					
Chloride % (Cl)					
Copper ppm (Cu)		6–9	10–50	>50	
Zinc ppm (Zn)		15–19	20–200	>200	
Manganese ppm (Mn)		30–49	50–300	>300	
Iron* ppm (Fe)		30–49	50–300	>300	
Boron ppm (B)		20–34	35–50	>50	
Molybdenum ppm (Mo)					

Araucaria bidwillii
(Bunya pine)

Plant part Most recently mature leaf — Ref: 9

Nutrient	Deficient	Low	Normal	High	Excess
Nitrogen % (N)		0.90–1.19	1.20–2.5	>2.5	
Phosphorus % (P)		0.12–0.15	0.16–0.3	>0.3	
Potassium % (K)		1.20–1.49	1.50–2.5	>2.5	
Calcium % (Ca)		0.50–0.69	0.70–1.5	>1.5	
Magnesium % (Mg)		0.15–0.19	0.20–0.5	>0.5	
Sulphur % (S)		0.09–0.12	0.13–0.25	>0.25	
Sodium % (Na)					
Chloride % (Cl)					
Copper ppm (Cu)		4–5	6–50	>50	
Zinc ppm (Zn)		12–19	20–200	>200	
Manganese ppm (Mn)		20–29	30–250	>250	
Iron* ppm (Fe)		40–49	50–300	>300	
Boron ppm (B)		10–14	15–40	>40	
Molybdenum ppm (Mo)					

Araucaria heterophylla
(Norfolk Island pine)

Plant part Most recently mature leaf — Ref: 7, 9

Nutrient[1]	Deficient	Low	Normal	High	Excess
Nitrogen % (N)		1.20–1.49	1.50–2.8	>2.8	
Phosphorus % (P)		0.15–0.19	0.20–0.3	>0.3	
Potassium % (K)		1.20–1.49	1.50–2.5	>2.5	
Calcium % (Ca)		0.50–0.69	0.70–1.5	>1.5	
Magnesium % (Mg)		0.15–0.19	0.20–0.5	>0.5	
Sulphur % (S)		0.10–0.14	0.15–0.25	>0.25	
Sodium % (Na)					
Chloride % (Cl)					
Copper ppm (Cu)		4–5	6–50	>50	
Zinc ppm (Zn)		20–24	25–200	>200	
Manganese ppm (Mn)		20–29	30–250	>250	
Iron* ppm (Fe)		40–49	50–300	>300	
Boron ppm (B)	<9	10–14	15–40	>40	>66
Molybdenum ppm (Mo)					

Asparagus densiflorus
(Asparagus fern, myers or sprengeri)

Plant part Most recently mature compound leaf — Ref: 3

Nutrient[1]	Deficient	Low	Normal	High	Excess
Nitrogen % (N)		1.20–1.49	1.50–2.5	>2.5	
Phosphorus % (P)		0.20–0.29	0.30–0.5	>0.5	
Potassium % (K)		1.40–1.99	2.0–3.0	>3.0	
Calcium % (Ca)		0.06–0.09	0.1–0.3	>0.3	
Magnesium % (Mg)		0.06–0.09	0.1–0.3	>0.3	
Sulphur % (S)		0.10–0.14	0.15–0.25	>0.25	
Sodium % (Na)					
Chloride % (Cl)					
Copper ppm (Cu)		6–9	10–50	>50	
Zinc ppm (Zn)		16–24	25–200	>200	
Manganese ppm (Mn)		30–39	40–300	>300	
Iron* ppm (Fe)		40–49	50–300	>300	
Boron ppm (B)		15–19	20–40	>40	
Molybdenum ppm (Mo)					

Asparagus retrofractus
(Asparagus fern)

Plant part Most recently mature compound leaf — Ref: 3

Nutrient[1]	Deficient	Low	Normal	High	Excess
Nitrogen % (N)		1.20–1.49	1.50–3.1	>3.1	
Phosphorus % (P)		0.13–0.19	0.20–0.3	>0.3	
Potassium % (K)		1.70–2.19	2.20–3.5	>3.5	
Calcium % (Ca)		0.25–0.39	0.40–0.6	>0.6	
Magnesium % (Mg)		0.06–0.09	0.10–0.3	>0.3	
Sulphur % (S)		0.11–0.14	0.15–0.3	>0.3	
Sodium % (Na)					
Chloride % (Cl)					
Copper ppm (Cu)		4–5	6–20	>20	
Zinc ppm (Zn)		16–24	25–200	>200	
Manganese ppm (Mn)		30–39	40–300	>300	
Iron* ppm (Fe)		40–49	50–300	>300	
Boron ppm (B)		15–19	20–40	>40	
Molybdenum ppm (Mo)					

Asplenium nidus
(Birdsnest fern)

Plant part Most recently mature frond — Ref: 9

Nutrient[1]	Deficient	Low	Normal	High	Excess
Nitrogen % (N)		1.60–2.09	2.10–3.2	>3.2	
Phosphorus % (P)		0.15–0.29	0.30–0.5	>0.5	
Potassium % (K)		2.00–2.49	2.50–4.2	>4.2	
Calcium % (Ca)		0.30–0.49	0.50–1.0	>1.0	
Magnesium % (Mg)		0.16–0.24	0.25–0.4	>0.4	
Sulphur % (S)		0.12–0.19	0.20–0.35	>0.35	
Sodium % (Na)					
Chloride % (Cl)					
Copper ppm (Cu)		5–7	8–20	>20	
Zinc ppm (Zn)		13–19	20–100	>100	
Manganese ppm (Mn)		20–29	30–300	>300	
Iron* ppm (Fe)		35–49	50–300	>300	
Boron ppm (B)		11–14	15–50	>50	
Molybdenum ppm (Mo)					

Aster novae-angliae
Aster novi-belgii

Plant part Most recently mature leaf
Growth stage Flower buds visible

Ref: 1

Nutrient[1]	Deficient	Low	Normal	High	Excess
Nitrogen % (N)			2.2–3.1		
Phosphorus % (P)			0.24–0.65		
Potassium % (K)			3.29–3.67		
Calcium % (Ca)			0.98–1.68		
Magnesium % (Mg)			0.18–0.35		
Sulphur % (S)					
Sodium % (Na)					
Chloride % (Cl)					
Copper ppm (Cu)					
Zinc ppm (Zn)			26–121		
Manganese ppm (Mn)			65–273		
Iron* ppm (Fe)			162–180		
Boron ppm (B)			37–46		
Molybdenum ppm (Mo)					

Azalea

Plant part Most recently mature leaf

Ref: 6

Nutrient[1]	Deficient	Low	Normal	High	Excess
Nitrogen % (N)			2.2–2.8		
Phosphorus % (P)			0.2–0.5		
Potassium % (K)			0.7–1.6		
Calcium % (Ca)			0.2–1.6		
Magnesium % (Mg)			0.1–0.6		
Sulphur % (S)					
Sodium % (Na)					
Chloride % (Cl)					
Copper ppm (Cu)			5–15		
Zinc ppm (Zn)			5–60		
Manganese ppm (Mn)			30–300		
Iron* ppm (Fe)			50–150		
Boron ppm (B)			15–100		
Molybdenum ppm (Mo)					

Banksia ericifolia

Plant part Hardened youngest fully expanded leaf — Ref: 17

Nutrient[1]	Deficient	Low	Normal	High	Excess
Nitrogen % (N)		<1.20	1.20–2.00		
Phosphorus % (P)		<0.17	0.17–0.18	>0.20	
Potassium % (K)		<0.82	0.82–1.00		
Calcium % (Ca)			0.57–0.65		
Magnesium % (Mg)			0.16–0.20		
Sulphur % (S)			0.35–0.40		
Sodium % (Na)			0.20–0.40		
Chloride % (Cl)			0.85–1.20		
Copper ppm (Cu)			<10		
Zinc ppm (Zn)			10–15		
Manganese ppm (Mn)			170–250		
Iron* ppm (Fe)			50–60		
Boron ppm (B)					
Molybdenum ppm (Mo)					

Banksia hookeriana

Plant part Hardened youngest fully expanded leaf — Ref: 17

Nutrient[1]	Deficient	Low	Normal	High	Excess
Nitrogen % (N)		<2.60	2.60–3.55		
Phosphorus % (P)		<0.09	0.09–0.32	>0.50	
Potassium % (K)		<0.80	0.80–1.20		
Calcium % (Ca)			0.58–1.00		
Magnesium % (Mg)			0.58–1.00		
Sulphur % (S)			0.29–0.33		
Sodium % (Na)			0.26–0.27		
Chloride % (Cl)			0.52–0.80		
Copper ppm (Cu)			7–8		
Zinc ppm (Zn)			21–47		
Manganese ppm (Mn)			345–500		
Iron* ppm (Fe)			68–111		
Boron ppm (B)					
Molybdenum ppm (Mo)					

Beaucarnea recurvata
(Pony tail palm)

Plant part Middle leaflet from most recent fully developed leaf — Ref: 9

Nutrient[1]	Deficient	Low	Normal	High	Excess
Nitrogen % (N)		<1.5	1.50–2.0	>2.0	
Phosphorus % (P)		<0.14	0.14–0.2	>0.2	
Potassium % (K)		<1.7	1.70–2.8	>2.8	
Calcium % (Ca)		<1.0	1.00–2.0	>2.0	
Magnesium % (Mg)		<0.2	0.20–0.3	>0.3	
Sulphur % (S)					
Sodium % (Na)					
Chloride % (Cl)					
Copper ppm (Cu)		<8	8–25	>25	
Zinc ppm (Zn)		<25	25–75	>75	
Manganese ppm (Mn)		<25	25–200	>200	
Iron* ppm (Fe)		<60	60–200	>200	
Boron ppm (B)		<20	20–35	>35	
Molybdenum ppm (Mo)					

Begonia hiemalis

Plant part Most recently mature leaf — Ref: 9

Nutrient[1]	Deficient	Low	Normal	High	Excess
Nitrogen % (N)		2.70–3.40	3.50–6.0	>6.0	
Phosphorus % (P)		0.20–0.29	0.30–0.75	>0.75	
Potassium % (K)		2.00–2.40	2.50–6.0	>6.0	
Calcium % (Ca)		0.60–0.90	1.00–2.5	>2.5	
Magnesium % (Mg)		0.25–0.29	0.30–0.7	>0.7	
Sulphur % (S)		0.20–0.29	0.30–0.7	>0.7	
Sodium % (Na)					
Chloride % (Cl)					
Copper ppm (Cu)		4–6	7–30	>30	
Zinc ppm (Zn)		20–24	25–200	>200	
Manganese ppm (Mn)		30–49	50–200	>200	
Iron* ppm (Fe)		40–49	50–200	>200	
Boron ppm (B)		15–19	20–75	>75	
Molybdenum ppm (Mo)					

Betula
(Birch tree)

Sampling time Current year's terminals at mid season
Plant part Most recently mature leaf

Ref: 2

Nutrient[1]	Deficient	Low	Normal	High	Excess
Nitrogen % (N)			2.50–4.00		
Phosphorus % (P)			0.15–0.30		
Potassium % (K)			1.00–1.50		
Calcium % (Ca)			0.30–1.50		
Magnesium % (Mg)			0.15–0.30		
Sulphur % (S)					
Sodium % (Na)					
Chloride % (Cl)					
Copper ppm (Cu)			6–12		
Zinc ppm (Zn)			15–50		
Manganese ppm (Mn)			30–100		
Iron* ppm (Fe)					
Boron ppm (B)			15–40		
Molybdenum ppm (Mo)			0.05–0.20		

Bougainvillea sp.
(Bougainvillea)

Plant part Most recently mature leaf

Ref: 9

Nutrient[1]	Deficient	Low	Normal	High	Excess
Nitrogen % (N)		2.00–2.40	2.50–4.50	>4.50	
Phosphorus % (P)		0.20–0.24	0.25–0.75	>0.75	
Potassium % (K)		2.50–2.90	3.00–5.50	>5.50	
Calcium % (Ca)		0.70–0.90	1.00–2.00	>2.00	
Magnesium % (Mg)		0.21–0.24	0.25–0.75	>0.75	
Sulphur % (S)		0.12–0.19	0.20–0.50	>0.50	
Sodium % (Na)					
Chloride % (Cl)					
Copper ppm (Cu)		6–7	8–50	>50	
Zinc ppm (Zn)		15–19	20–200	>200	
Manganese ppm (Mn)		30–49	50–200	>200	
Iron* ppm (Fe)		40–49	50–300	>300	
Boron ppm (B)		20–24	25–75	>75	
Molybdenum ppm (Mo)					

Bucida buceras
(Black olive)

Plant part Most recently mature leaf Ref: 9

Nutrient[1]	Deficient	Low	Normal	High	Excess
Nitrogen % (N)		<1.60	1.60–3.00	>3.00	
Phosphorus % (P)		<0.15	0.15–0.75	>0.75	
Potassium % (K)		<0.70	0.70–3.50	>3.50	
Calcium % (Ca)		<0.25	0.25–1.00	>1.00	
Magnesium % (Mg)		<0.25	0.25–1.00	>1.00	
Sulphur % (S)		<0.20	0.20–0.75	>0.75	
Sodium % (Na)					
Chloride % (Cl)					
Copper ppm (Cu)		<5	5–25	>25	
Zinc ppm (Zn)		<0	20–100	>100	
Manganese ppm (Mn)		<40	40–200	>200	
Iron* ppm (Fe)		<50	50–200	>200	
Boron ppm (B)		<25	25–75	>75	
Molybdenum ppm (Mo)					

Buxus macrophylla var. _japonica_
(Japanese boxwood)

Plant part Most recently mature leaf Ref: 9

Nutrient[1]	Deficient	Low	Normal	High	Excess
Nitrogen % (N)		<3.0	3.00–3.6	>3.6	
Phosphorus % (P)		<0.30	0.30–0.5	>0.5	
Potassium % (K)		<1.25	1.25–2.0	>2.0	
Calcium % (Ca)		<1.0	1.00–2.0	>2.0	
Magnesium % (Mg)		<0.30	0.30–0.6	>0.6	
Sulphur % (S)					
Sodium % (Na)					
Chloride % (Cl)					
Copper ppm (Cu)					
Zinc ppm (Zn)					
Manganese ppm (Mn)					
Iron* ppm (Fe)					
Boron ppm (B)					
Molybdenum ppm (Mo)					

Cactus
(Holiday)

Plant part Most recently mature leaf — Ref: 6

Nutrient[1]	Deficient	Low	Normal	High	Excess
Nitrogen % (N)			2.7–3.7		
Phosphorus % (P)			0.5–0.9		
Potassium % (K)			6.2–7.1		
Calcium % (Ca)			0.7–0.9		
Magnesium % (Mg)			1.6–2.2		
Sulphur % (S)					
Sodium % (Na)					
Chloride % (Cl)					
Copper ppm (Cu)			10–15		
Zinc ppm (Zn)			50–65		
Manganese ppm (Mn)			35–130		
Iron* ppm (Fe)			105–110		
Boron ppm (B)			65–70		
Molybdenum ppm (Mo)					

Caladium
(Caladium)

Plant part Unfurled leaves with midribs — Ref: 9

Nutrient[1]	Deficient	Low	Normal	High	Excess
Nitrogen % (N)		2.00–3.5	3.6–4.5	>4.8	
Phosphorus % (P)		0.10–0.2	0.3–0.7	>0.7	
Potassium % (K)		1.50–2.2	2.3–4.0	>4.0	
Calcium % (Ca)		<1.0	1.0–1.5	>1.5	
Magnesium % (Mg)		<0.2	0.2–0.4	>0.4	
Sulphur % (S)					
Sodium % (Na)					
Chloride % (Cl)					
Copper ppm (Cu)		<6	6–10	>10	
Zinc ppm (Zn)		<30	30–150	>150	
Manganese ppm (Mn)		<50	50–210	>210	
Iron* ppm (Fe)		<60	60–100	>100	
Boron ppm (B)		<40	40–100	>100	
Molybdenum ppm (Mo)					

Calathea sp.
(Calathea)

Plant part Most recently mature leaf — Ref: 4, 9

Nutrient[1]	Deficient	Low	Normal	High	Excess
Nitrogen % (N)		2.00–2.49	2.50–4.0	>4.0	
Phosphorus % (P)		0.15–0.19	0.20–0.5	>0.5	
Potassium % (K)		2.00–2.49	2.50–4.5	>4.5	
Calcium % (Ca)		0.25–0.49	0.50–1.5	>1.5	
Magnesium % (Mg)		0.20–0.24	0.25–0.6	>0.6	
Sulphur % (S)					
Sodium % (Na)					
Chloride % (Cl)					
Copper ppm (Cu)		4–5	6–50	>50	
Zinc ppm (Zn)		15–19	20–200	>200	
Manganese ppm (Mn)		21–29	30–200	>200	
Iron* ppm (Fe)		25–29	30–200	>200	
Boron ppm (B)		12–17	18–50	>50	
Molybdenum ppm (Mo)					

Calla

Plant part Most recently mature leaf — Ref: 6

Nutrient[1]	Deficient	Low	Normal	High	Excess
Nitrogen % (N)			2.9–3.0		
Phosphorus % (P)			0.3–0.4		
Potassium % (K)			3.9–4.4		
Calcium % (Ca)			0.9–1.1		
Magnesium % (Mg)			0.3–0.4		
Sulphur % (S)					
Sodium % (Na)					
Chloride % (Cl)					
Copper ppm (Cu)			5–10		
Zinc ppm (Zn)			30–45		
Manganese ppm (Mn)			635–690		
Iron* ppm (Fe)			95–130		
Boron ppm (B)			30–40		
Molybdenum ppm (Mo)					

Camellia sinensis
(Tea plant)

Sampling time Mid season
Plant part Most recently mature leaf

Ref: 2

Nutrient[1]	Deficient	Low	Normal	High	Excess
Nitrogen % (N)			4.50–5.20		
Phosphorus % (P)			0.35–0.60		
Potassium % (K)			1.60–2.30		
Calcium % (Ca)			0.40–0.80		
Magnesium % (Mg)			0.20–0.40		
Sulphur % (S)					
Sodium % (Na)					
Chloride % (Cl)					
Copper ppm (Cu)			7–15		
Zinc ppm (Zn)			30–80		
Manganese ppm (Mn)			100–500		
Iron* ppm (Fe)					
Boron ppm (B)			30–50		
Molybdenum ppm (Mo)			0.20–0.50		

Carissa grandiflora
(Carissa or Natal plum)

Plant part Most recent fully developed leaf

Ref: 7, 9

Nutrient[1]	Deficient	Low	Normal	High	Excess
Nitrogen % (N)		1.50–1.79	1.80–3.5	>3.5	
Phosphorus % (P)		0.15–0.17	0.18–0.6	>0.6	
Potassium % (K)		1.20–1.49	1.50–3.5	>3.5	
Calcium % (Ca)		0.80–0.99	1.00–3.0	>3.0	
Magnesium % (Mg)		0.22–0.24	0.25–1.0	>1.0	
Sulphur % (S)		0.13–0.19	0.20–0.4	>0.4	
Sodium % (Na)					
Chloride % (Cl)					
Copper ppm (Cu)		4–5	6–50	> 50	
Zinc ppm (Zn)		16–19	20–200	>200	
Manganese ppm (Mn)		40–49	50–250	>250	
Iron* ppm (Fe)		40–49	50–200	>200	
Boron ppm (B)	<19	20–24	25–100	>100	>126
Molybdenum ppm (Mo)					

Caryopteris incana
(Blue spirea)

Plant part Most recently mature leaf
Growth stage Flower buds visible

Ref: 1

Nutrient[1]	Deficient	Low	Normal	High	Excess
Nitrogen % (N)			3.0		
Phosphorus % (P)			0.21		
Potassium % (K)			1.47		
Calcium % (Ca)			1.00		
Magnesium % (Mg)			0.16		
Sulphur % (S)					
Sodium % (Na)					
Chloride % (Cl)					
Copper ppm (Cu)					
Zinc ppm (Zn)			57		
Manganese ppm (Mn)			80		
Iron* ppm (Fe)			125		
Boron ppm (B)			15		
Molybdenum ppm (Mo)					

Cattelya sp.
(Cattelya orchid)

Plant part Most recently mature leaf

Ref: 9

Nutrient[1]	Deficient	Low	Normal	High	Excess
Nitrogen % (N)		1.20–1.49	1.50–2.50	>2.50	
Phosphorus % (P)		0.10–0.12	0.13–0.75	>0.75	
Potassium % (K)		1.50–1.99	2.00–3.50	>3.50	
Calcium % (Ca)		0.35–0.49	0.50–2.00	>2.00	
Magnesium % (Mg)		0.20–0.29	0.30–0.70	>0.70	
Sulphur % (S)		0.12–0.14	0.15–0.75	>0.75	
Sodium % (Na)					
Chloride % (Cl)					
Copper ppm (Cu)		2–4	5–20	>20	
Zinc ppm (Zn)		20–24	25–200	>200	
Manganese ppm (Mn)		30–39	40–200	>200	
Iron* ppm (Fe)		40–49	50–200	>200	
Boron ppm (B)		20–24	25–75	>75	
Molybdenum ppm (Mo)					

Celosia argentea
(Cockscomb)

Plant part Most recently mature leaf
Growth stage Flower buds visible
Ref: 1

Nutrient[1]	Deficient	Low	Normal	High	Excess
Nitrogen % (N)			3.9		
Phosphorus % (P)			0.43		
Potassium % (K)			5.06		
Calcium % (Ca)			2.86		
Magnesium % (Mg)			1.36		
Sulphur % (S)					
Sodium % (Na)					
Chloride % (Cl)					
Copper ppm (Cu)					
Zinc ppm (Zn)			182		
Manganese ppm (Mn)			261		
Iron* ppm (Fe)			189		
Boron ppm (B)			23		
Molybdenum ppm (Mo)					

Chamaedorea elegans*, *C. erumpens
(Bamboo parlor palm)

Plant part Middle leaflets from most recently mature frond
Ref: 3

Nutrient[1]	Deficient	Low	Normal	High	Excess
Nitrogen % (N)	<1.90	2.0–2.40	2.50–3.50	3.60–4.50	>4.50
Phosphorus % (P)	<0.10	0.11–0.14	0.15–0.30	0.31–0.75	>0.76
Potassium % (K)	<1.20	1.25–1.55	1.60–2.75	2.80–4.00	>4.05
Calcium % (Ca)	<0.39	0.40–0.99	1.00–2.50	2.51–3.25	>3.26
Magnesium % (Mg)	<0.20	0.21–0.24	0.25–0.75	0.76–1.00	>1.01
Sulphur % (S)	<0.14	0.15–0.20	0.21–0.40	0.41–0.75	>0.76
Sodium % (Na)			0–0.20	0.21–0.50	>0.51
Chloride % (Cl)					
Copper ppm (Cu)	<3	4–5	6–50	51–200	>201
Zinc ppm (Zn)	<17	18–24	25–200	201–500	>501
Manganese ppm (Mn)	<39	40–49	50–250	251–1000	>1001
Iron* ppm (Fe)	<39	40–49	50–300	301–1000	>1001
Boron ppm (B)	<17	18–24	25–60	61–100	>101
Molybdenum ppm (Mo)					

***Chamelaucium* spp.**
(Waxflower)

Plant part Top 2.5–4 mm of actively growing stems — Ref: 21

Nutrient[1]	Deficient	Low	Normal	High	Excess
Nitrogen % (N)			1.5–2.5	>4.0	
Phosphorus % (P)			0.1–0.2		
Potassium % (K)			0.8–1.5		
Calcium % (Ca)			0.4		
Magnesium % (Mg)			0.3		
Sulphur % (S)					
Sodium % (Na)			0.2		
Chloride % (Cl)					
Copper ppm (Cu)			5		
Zinc ppm (Zn)			30		
Manganese ppm (Mn)			30		
Iron* ppm (Fe)			80		
Boron ppm (B)					
Molybdenum ppm (Mo)					

Chlorophytum comosum
(Spider plant)

Plant part Most recently mature leaf — Ref: 9

Nutrient[1]	Deficient	Low	Normal	High	Excess
Nitrogen % (N)		1.20–1.69	1.70–3.0	>3.0	
Phosphorus % (P)		0.11–0.14	0.15–0.4	>0.4	
Potassium % (K)		2.00–2.49	2.50–5.0	>5.0	
Calcium % (Ca)		0.80–0.99	1.00–2.5	>2.5	
Magnesium % (Mg)		0.20–0.24	0.25–1.5	>1.5	
Sulphur % (S)					
Sodium % (Na)					
Chloride % (Cl)					
Copper ppm (Cu)		<8	8–25	>25	
Zinc ppm (Zn)		<25	25–200	>200	
Manganese ppm (Mn)		<50	50–75	>75	
Iron* ppm (Fe)		<60	60–150	>150	
Boron ppm (B)		<25	25–40	>40	
Molybdenum ppm (Mo)					

Chrysalidocarpus lutescens
(Areca palm)

Plant part Middle leaflets from most recently mature frond — Ref: 3

Nutrient[1]	Deficient	Low	Normal	High	Excess
Nitrogen % (N)	<1.90	2.0–2.40	2.50–3.50	3.60–4.50	>4.50
Phosphorus % (P)	<0.10	0.11–0.14	0.15–0.30	0.31–0.75	>0.76
Potassium % (K)	<1.20	1.25–1.55	1.60–2.75	2.80–4.00	>4.05
Calcium % (Ca)	<0.39	0.40–0.99	1.00–2.50	2.51–3.25	>3.26
Magnesium % (Mg)	<0.20	0.21–0.24	0.25–0.75	0.76–1.00	>1.01
Sulphur % (S)	<0.14	0.15–0.20	0.21–0.40	0.41–0.75	>0.76
Sodium % (Na)			0–0.20	0.21–0.50	>0.51
Chloride % (Cl)					
Copper ppm (Cu)	<3	4–5	6–50	51–200	>201
Zinc ppm (Zn)	<17	18–24	25–200	201–500	>501
Manganese ppm (Mn)	<39	40–49	50–250	251–1000	>1001
Iron* ppm (Fe)	<39	40–49	50–300	301–1000	>1001
Boron ppm (B)	<17	18–24	25–60	61–100	>101
Molybdenum ppm (Mo)					

Chrysanthemum morifolium
(mum or pompom)

Sampling time To bud start
Plant part 4th leaf from tip – omit unfurled leaves — Ref: 9

Nutrient[1]	Deficient	Low	Normal	High	Excess
Nitrogen % (N)		3.80–3.90	4.00–6.0	>6.0	
Phosphorus % (P)		0.22–0.24	0.25–1.0	>1.0	
Potassium % (K)		3.60–3.90	4.00–6.00	>6.0	
Calcium % (Ca)		0.70–0.90	1.00–2.0	>2.0	
Magnesium % (Mg)		0.20–0.24	0.25–1.0	>1.0	
Sulphur % (S)		0.20–0.24	0.25–0.7	>0.7	
Sodium % (Na)					
Chloride % (Cl)					
Copper ppm (Cu)		4–5	6–30	>30	
Zinc ppm (Zn)		18–19	20–250	>250	
Manganese ppm (Mn)		30–49	50–250	>250	
Iron* ppm (Fe)		40–49	50–250	>250	
Boron ppm (B)		21–24	25–75	>75	
Molybdenum ppm (Mo)					

Chrysanthemum morifolium
(mum or pompom)

Sampling time Bud start to harvest
Plant part 4th leaf from tip – omit unfurled leaves

Ref: 9

Nutrient[1]	Deficient	Low	Normal	High	Excess
Nitrogen % (N)		3.00–3.40	3.50–5.00	>5.00	
Phosphorus % (P)		0.20–0.22	0.23–0.7	>0.7	
Potassium % (K)		3.00–3.40	3.50–5.0	>5.0	
Calcium % (Ca)		0.90–1.01	1.20–2.5	>2.5	
Magnesium % (Mg)		0.20–0.24	0.25–1.0	>1.0	
Sulphur % (S)		0.20–0.24	0.25–0.7	>0.7	
Sodium % (Na)					
Chloride % (Cl)					
Copper ppm (Cu)		4–5	6–30	>30	
Zinc ppm (Zn)		18–19	20–250	>250	
Manganese ppm (Mn)		39–40	50–250	>250	
Iron* ppm (Fe)		40–49	50–250	>250	
Boron ppm (B)		21–24	25–75	>75	
Molybdenum ppm (Mo)					

Chrysobalanus lcaco
(Coco plum)

Plant part Most recently mature leaf

Ref: 9

Nutrient[1]	Deficient	Low	Normal	High	Excess
Nitrogen % (N)		1.50–1.90	2.00–3.00	>3.0	
Phosphorus % (P)		0.20–0.24	0.25–1.0	>1.0	
Potassium % (K)		0.70–0.90	1.00–2.5	>2.5	
Calcium % (Ca)		0.50–0.70	0.80–2.0	>2.0	
Magnesium % (Mg)		0.21–0.24	0.25–1.0	>1.0	
Sulphur % (S)		0.13–0.19	0.20–0.4	>0.4	
Sodium % (Na)					
Chloride % (Cl)					
Copper ppm (Cu)		6–9	10–50	>50	
Zinc ppm (Zn)		15–19	20–200	>200	
Manganese ppm (Mn)		25–39	40–200	>200	
Iron* ppm (Fe)		40–49	50–200	>200	
Boron ppm (B)		20–24	25–100	>100	
Molybdenum ppm (Mo)					

Cirsium japonicum
(Japanese thistle)

Plant part Most recently mature leaf
Growth stage Flower buds visible

Ref: 1

Nutrient[1]	Deficient	Low	Normal	High	Excess
Nitrogen % (N)			2.6		
Phosphorus % (P)			0.19		
Potassium % (K)			2.24		
Calcium % (Ca)			0.60		
Magnesium % (Mg)			0.29		
Sulphur % (S)					
Sodium % (Na)					
Chloride % (Cl)					
Copper ppm (Cu)					
Zinc ppm (Zn)			21		
Manganese ppm (Mn)			130		
Iron* ppm (Fe)			462		
Boron ppm (B)			14		
Molybdenum ppm (Mo)					

Clarkia or *Godetia* spp.

Plant part Most recently mature leaf
Growth stage Flower buds visible

Ref: 1

Nutrient[1]	Deficient	Low	Normal	High	Excess
Nitrogen % (N)			3.2		
Phosphorus % (P)			0.26		
Potassium % (K)			4.18		
Calcium % (Ca)			1.03		
Magnesium % (Mg)			0.34		
Sulphur % (S)					
Sodium % (Na)					
Chloride % (Cl)					
Copper ppm (Cu)					
Zinc ppm (Zn)			36		
Manganese ppm (Mn)			239		
Iron* ppm (Fe)			189		
Boron ppm (B)			23		
Molybdenum ppm (Mo)					

Cocos nucifera
(Coconut palm)

Sampling time Current year's terminals at mid season
Plant part Most recently mature leaf Ref: 2

Nutrient[1]	Deficient	Low	Normal	High	Excess
Nitrogen % (N)			1.87–2.50		
Phosphorus % (P)			0.16–0.20		
Potassium % (K)			1.45–3.00		
Calcium % (Ca)			0.25–0.30		
Magnesium % (Mg)			0.16–0.32		
Sulphur % (S)			0.18–0.22		
Sodium % (Na)					
Chloride % (Cl)					
Copper ppm (Cu)					
Zinc ppm (Zn)					
Manganese ppm (Mn)					
Iron* ppm (Fe)					
Boron ppm (B)					
Molybdenum ppm (Mo)					

Codiaeum variegatum
(Croton)

Plant part Most recently mature leaf Ref: 9

Nutrient[1]	Deficient	Low	Normal	High	Excess
Nitrogen % (N)		1.20–1.49	1.50–3.0	>3.0	
Phosphorus % (P)		0.20–0.24	0.25–0.5	>0.5	
Potassium % (K)		1.00–1.29	1.30–3.0	>3.0	
Calcium % (Ca)		0.70–0.99	1.00–2.5	>2.5	
Magnesium % (Mg)		0.25–0.29	0.30–1.0	>1.0	
Sulphur % (S)		0.13–0.19	0.20–0.4	>0.4	
Sodium % (Na)					
Chloride % (Cl)					
Copper ppm (Cu)		7–9	10–50	>50	
Zinc ppm (Zn)		15–19	20–250	>250	
Manganese ppm (Mn)		40–49	50–200	>200	
Iron* ppm (Fe)		40–49	50–200	>200	
Boron ppm (B)		20–24	25–75	>75	
Molybdenum ppm (Mo)					

Coffea arabica
(Coffee)

Plant part Most recently mature leaf — Ref: 9

Nutrient[1]	Deficient	Low	Normal	High	Excess
Nitrogen % (N)		1.60–2.49	2.50–3.5	>3.5	
Phosphorus % (P)		0.11–0.14	0.15–0.35	>0.35	
Potassium % (K)		1.50–1.99	2.00–3.00	>3.0	
Calcium % (Ca)		0.40–0.79	0.80–1.60	>1.6	
Magnesium % (Mg)		0.16–0.29	0.30–0.50	>0.5	
Sulphur % (S)		0.16–0.24	0.25–0.50	>0.5	
Sodium % (Na)					
Chloride % (Cl)					
Copper ppm (Cu)		6–9	10–50	>50	
Zinc ppm (Zn)		11–14	15–200	>200	
Manganese ppm (Mn)		35–49	50–300	>300	
Iron* ppm (Fe)		70–89	90–300	>300	
Boron ppm (B)		20–24	25–75	>75	
Molybdenum ppm (Mo)					

Conospermum caeruleum

Plant part Hardened youngest fully expanded leaf — Ref: 17

Nutrient[1]	Deficient	Low	Normal	High	Excess
Nitrogen % (N)		<1.60	1.60–1.80		
Phosphorus % (P)			0.10–0.30	>0.40	
Potassium % (K)			0.60–0.80		
Calcium % (Ca)			0.66–0.80		
Magnesium % (Mg)			0.13–0.17		
Sulphur % (S)			0.16–0.18		
Sodium % (Na)			0.23–0.25		
Chloride % (Cl)			0.35–0.39		
Copper ppm (Cu)			10		
Zinc ppm (Zn)			165		
Manganese ppm (Mn)			275–340		
Iron* ppm (Fe)			40–170		
Boron ppm (B)					
Molybdenum ppm (Mo)					

Conospermum eatoniae

Plant part Hardened youngest fully mature leaf — Ref: 17

Nutrient[1]	Deficient	Low	Normal	High	Excess
Nitrogen % (N)		<1.20	1.20–1.55		
Phosphorus % (P)			0.10–0.20	>0.30	
Potassium % (K)			0.35–0.40		
Calcium % (Ca)			1.10–1.30		
Magnesium % (Mg)			0.35–0.40		
Sulphur % (S)			0.55–0.60		
Sodium % (Na)			0.35–0.40		
Chloride % (Cl)			0.35–0.40		
Copper ppm (Cu)					
Zinc ppm (Zn)					
Manganese ppm (Mn)					
Iron* ppm (Fe)					
Boron ppm (B)					
Molybdenum ppm (Mo)					

Cornus racemosa* and *Cornus alba
(Dogwood)

Plant part Most fully mature leaf — Ref: 9

Nutrient[1]	Deficient	Low	Normal	High	Excess
Nitrogen % (N)			1.82–2.6		
Phosphorus % (P)			0.31–0.62		
Potassium % (K)			1.10–1.18		
Calcium % (Ca)			2.23–2.95		
Magnesium % (Mg)			0.36–0.58		
Sulphur % (S)					
Sodium % (Na)					
Chloride % (Cl)					
Copper ppm (Cu)			9–10		
Zinc ppm (Zn)			22–33		
Manganese ppm (Mn)			10–13		
Iron* ppm (Fe)			75–82		
Boron ppm (B)			21–34		
Molybdenum ppm (Mo)			1.8–4.8		

Cotoneaster
(Cotoneaster apiculata)

Plant part Most recently mature leaf — Ref: 9

Nutrient[1]	Deficient	Low	Normal	High	Excess
Nitrogen % (N)			2.80		
Phosphorus % (P)			0.34		
Potassium % (K)			2.00		
Calcium % (Ca)			1.10		
Magnesium % (Mg)			0.27		
Sulphur % (S)					
Sodium % (Na)					
Chloride % (Cl)					
Copper ppm (Cu)			14		
Zinc ppm (Zn)			43		
Manganese ppm (Mn)			218		
Iron* ppm (Fe)			228		
Boron ppm (B)			30		
Molybdenum ppm (Mo)			0.65		

Crossandra infundibuliformis

Plant part Most recently mature leaf — Ref: 3

Nutrient[1]	Deficient	Low	Normal	High	Excess
Nitrogen % (N)			3.0–4.0		
Phosphorus % (P)			0.25–0.4		
Potassium % (K)			3.0–4.0		
Calcium % (Ca)			1.2–1.6		
Magnesium % (Mg)			0.4–0.6		
Sulphur % (S)					
Sodium % (Na)					
Chloride % (Cl)					
Copper ppm (Cu)			8–60		
Zinc ppm (Zn)			25–250		
Manganese ppm (Mn)			50–300		
Iron* ppm (Fe)			50–300		
Boron ppm (B)					
Molybdenum ppm (Mo)					

Cyclamen persicum
(Cyclamen or sowbread)

Sampling time At onset of flowering
Plant part Most recently mature leaf — Ref: 2

Nutrient[1]	Deficient	Low	Normal	High	Excess
Nitrogen % (N)			2.40–3.40		
Phosphorus % (P)			0.25–0.40		
Potassium % (K)			2.60–4.50		
Calcium % (Ca)			0.80–1.20		
Magnesium % (Mg)			0.25–0.50		
Sulphur % (S)					
Sodium % (Na)					
Chloride % (Cl)					
Copper ppm (Cu)			5–12		
Zinc ppm (Zn)			20–60		
Manganese ppm (Mn)			30–100		
Iron* ppm (Fe)					
Boron ppm (B)			25–60		
Molybdenum ppm (Mo)			0.15–0.40		

Cymbidium sp.
(Cymbidium orchid)

Plant part Most recently mature leaf — Ref: 9

Nutrient[1]	Deficient	Low	Normal	High	Excess
Nitrogen % (N)		1.20–1.49	1.50–2.50	>2.50	
Phosphorus % (P)		0.10–0.12	0.13–0.75	>0.75	
Potassium % (K)		1.50–1.99	2.00–3.50	>3.50	
Calcium % (Ca)		0.35–0.49	0.50–2.00	>2.00	
Magnesium % (Mg)		0.20–0.29	0.30–0.70	>0.70	
Sulphur % (S)		0.12–0.14	0.15–0.75	>0.75	
Sodium % (Na)					
Chloride % (Cl)					
Copper ppm (Cu)		2–4	5–20	>20	
Zinc ppm (Zn)		20–24	25–200	>200	
Manganese ppm (Mn)		30–39	40–200	>200	
Iron* ppm (Fe)		40–49	50–200	>200	
Boron ppm (B)		20–24	25–75	>75	
Molybdenum ppm (Mo)					

Cypripedium sp.
(Lady-slipper or cypripedium orchid)

Plant part Most recently mature leaf

Ref: 9

Nutrient[1]	Deficient	Low	Normal	High	Excess
Nitrogen % (N)		1.80–2.29	2.30–3.5	>3.5	
Phosphorus % (P)		0.15–0.19	0.20–0.7	>0.7	
Potassium % (K)		1.50–1.99	2.0–3.5	>3.5	
Calcium % (Ca)		0.50–0.74	0.75–2.0	>2.0	
Magnesium % (Mg)		0.15–0.19	0.20–0.7	>0.7	
Sulphur % (S)		0.15–0.19	0.20–0.7	>0.7	
Sodium % (Na)					
Chloride % (Cl)					
Copper ppm (Cu)		2–4	5–20	>20	
Zinc ppm (Zn)		20–24	25–200	>200	
Manganese ppm (Mn)		40–49	50–200	>200	
Iron* ppm (Fe)		40–49	50–200	>200	
Boron ppm (B)		20–24	25–75	>75	
Molybdenum ppm (Mo)					

Delphinium hybrids

Plant part Most recently mature leaf
Growth stage Flower buds visible

Ref: 1

Nutrient[1]	Deficient	Low	Normal	High	Excess
Nitrogen % (N)			3.2		
Phosphorus % (P)			0.33		
Potassium % (K)			3.52		
Calcium % (Ca)			2.86		
Magnesium % (Mg)			0.70		
Sulphur % (S)					
Sodium % (Na)					
Chloride % (Cl)					
Copper ppm (Cu)					
Zinc ppm (Zn)			35		
Manganese ppm (Mn)			59		
Iron* ppm (Fe)			617		
Boron ppm (B)			18		
Molybdenum ppm (Mo)					

Dianthus caryophyllus
(Carnation)

Plant part 4th and 5th leaf pair from growth tip (15 cm)
Growth stage Newly planted. Prior to flowering. Ref: 2

Nutrient[1]	Deficient	Low	Normal	High	Excess
Nitrogen % (N)			3.0–5.0 2.80–4.20		
Phosphorus % (P)			0.1–0.5 0.25–0.45		
Potassium % (K)			2.0–6.0 2.50–5.00		
Calcium % (Ca)			0.6–2.0 1.00–2.00		
Magnesium % (Mg)			0.2–0.6 0.25–0.50		
Sulphur % (S)					
Sodium % (Na)					
Chloride % (Cl)					
Copper ppm (Cu)			5–30 8–15		
Zinc ppm (Zn)			15–75 25–80		
Manganese ppm (Mn)			30–445 40–120		
Iron* ppm (Fe)			30–150		
Boron ppm (B)			20–400 30–80		
Molybdenum ppm (Mo)			0.25–1.00		

Dianthus caryophyllus
(Carnation)

Plant part 5th and 6th leaf pairs from non-flowering shoots Ref: 7, 9

Nutrient[1]	Deficient	Low	Normal	High	Excess
Nitrogen % (N)		2.90–3.19	3.20–5.2	>5.2	
Phosphorus % (P)		0.20–0.24	0.25–0.8	>0.8	
Potassium % (K)		2.00–2.79	2.80–6.0	>6.0	
Calcium % (Ca)		0.60–0.99	1.00–2.0	>2.0	
Magnesium % (Mg)		0.15–0.24	0.25–0.7	>0.7	
Sulphur % (S)		0.20–0.24	0.25–0.8	>0.8	
Sodium % (Na)					
Chloride % (Cl)					
Copper ppm (Cu)		5–7	8–30	>30	
Zinc ppm (Zn)		16–24	25–200	>200	
Manganese ppm (Mn)		31–49	50–200	>200	
Iron* ppm (Fe)		31–49	50–200	>200	
Boron ppm (B)	<24	25–29	30–100	101–135	>135
Molybdenum ppm (Mo)					

Dieffenbachia
(Dumb cane)

Plant part Most recently mature leaf minus petiole — Ref: 9

Nutrient[1]	Deficient	Low	Normal	High	Excess
Nitrogen % (N)		2.30–2.69	2.7–3.5	>3.5	
Phosphorus % (P)		0.15–0.19	0.2–0.5	>0.5	
Potassium % (K)		2.50–3.49	3.5–4.5	>4.5	
Calcium % (Ca)		0.70–0.99	1.0–2.0	>2.0	
Magnesium % (Mg)		0.20–0.29	0.3–0.75	>0.75	
Sulphur % (S)		0.15–0.19	0.2–0.5	>0.5	
Sodium % (Na)					
Chloride % (Cl)					
Copper ppm (Cu)		5–7	8–50	>50	
Zinc ppm (Zn)		15–19	20–200	>200	
Manganese ppm (Mn)		35–49	50–300	>300	
Iron* ppm (Fe)		40–49	50–300	>300	
Boron ppm (B)		12–14	15–50	>50	
Molybdenum ppm (Mo)					

Dizygotheca elegantissima
(False aralia)

Plant part Most recently mature leaf — Ref: 9

Nutrient[1]	Deficient	Low	Normal	High	Excess
Nitrogen % (N)		2.20–2.49	2.50–3.5	>3.5	
Phosphorus % (P)		0.18–0.24	0.25–0.6	>0.6	
Potassium % (K)		1.40–1.79	1.80–3.5	>3.5	
Calcium % (Ca)		0.30–0.49	0.50–2.0	>2.0	
Magnesium % (Mg)		0.20–0.24	0.25–0.4	>0.4	
Sulphur % (S)		0.18–0.24	0.25–0.5	>0.5	
Sodium % (Na)					
Chloride % (Cl)					
Copper ppm (Cu)		4–5	6–50	>50	
Zinc ppm (Zn)		16–19	20–200	>200	
Manganese ppm (Mn)		40–49	50–300	>300	
Iron* ppm (Fe)		<50	50–300	>300	
Boron ppm (B)		20–24	25–50	>50	
Molybdenum ppm (Mo)					

Dracaena

Plant part Most recently mature leaf — Ref: 4, 7

Nutrient[1]	Deficient	Low	Normal	High	Excess
Nitrogen % (N)			2.0–3.0		
Phosphorus % (P)			0.2–0.3		
Potassium % (K)			3.0–4.0		
Calcium % (Ca)			1.5–2.0		
Magnesium % (Mg)			0.3–0.6		
Sulphur % (S)					
Sodium % (Na)					
Chloride % (Cl)					
Copper ppm (Cu)			8–60		
Zinc ppm (Zn)			25–250		
Manganese ppm (Mn)			50–300		
Iron* ppm (Fe)			50–300		
Boron ppm (B)	<10		16–50		>101
Molybdenum ppm (Mo)					

Elaeis guineensis
(African Oil palm)

Plant part Middle leaflets from most recently mature frond minus petiole — Ref: 3

Nutrient[1]	Deficient	Low	Normal	High	Excess
Nitrogen % (N)	<1.90	2.00–2.30	2.40–3.00		
Phosphorus % (P)	<0.11	0.12–0.14	0.15–0.25		
Potassium % (K)	<0.74	0.75–0.99	1.00–1.80		
Calcium % (Ca)	<0.29	0.30–0.59	0.60–1.00		
Magnesium % (Mg)	<0.19	0.20–0.23	0.24–0.35		
Sulphur % (S)					
Sodium % (Na)			0–0.20		
Chloride % (Cl)					
Copper ppm (Cu)	<2	3	4–20		
Zinc ppm (Zn)	<9	10–13	14–100		
Manganese ppm (Mn)	<99	100–149	150–300		
Iron* ppm (Fe)	<29	30–39	40–100		
Boron ppm (B)	<4	5–7	8–30		
Molybdenum ppm (Mo)			1–10		

Epipremnum aureum

Plant part Most recently mature leaf — Ref: 7, 19

Nutrient[1]	Deficient	Low	Normal	High	Excess
Nitrogen % (N)			2.5–3.5		
Phosphorus % (P)			0.2–0.35		
Potassium % (K)			3.0–4.5		
Calcium % (Ca)			1.0–1.5		
Magnesium % (Mg)			0.3–0.6		
Sulphur % (S)					
Sodium % (Na)					
Chloride % (Cl)					
Copper ppm (Cu)			8–60		
Zinc ppm (Zn)			25–250		
Manganese ppm (Mn)			50–300		
Iron* ppm (Fe)			50–300		
Boron ppm (B)	<14		20–50		>76
Molybdenum ppm (Mo)					

Eryngium planum

Plant part Most recently mature leaf
Growth stage Flower buds visible — Ref: 1

Nutrient[1]	Deficient	Low	Normal	High	Excess
Nitrogen % (N)			4.05		
Phosphorus % (P)			0.63		
Potassium % (K)			3.36		
Calcium % (Ca)			1.19		
Magnesium % (Mg)			0.47		
Sulphur % (S)					
Sodium % (Na)					
Chloride % (Cl)					
Copper ppm (Cu)					
Zinc ppm (Zn)			61		
Manganese ppm (Mn)			84		
Iron* ppm (Fe)			257		
Boron ppm (B)			28		
Molybdenum ppm (Mo)					

Eugenia
(Eugenia)

Plant part Most recently mature leaf — Ref: 9

Nutrient[1]	Deficient	Low	Normal	High	Excess
Nitrogen % (N)		1.20–1.49	1.5–2.5	>2.5	
Phosphorus % (P)		0.21–0.39	0.4–0.8	>0.8	
Potassium % (K)		1.01–1.49	1.5–3.0	>3.0	
Calcium % (Ca)		0.70–0.99	1.0–2.5	>2.5	
Magnesium % (Mg)		0.15–0.19	0.2–0.8	>0.8	
Sulphur % (S)		0.13–0.19	0.2–0.4	>0.4	
Sodium % (Na)					
Chloride % (Cl)					
Copper ppm (Cu)		6–9	10–50	>50	
Zinc ppm (Zn)		16–19	20–200	>200	
Manganese ppm (Mn)		25–49	50–200	>200	
Iron* ppm (Fe)		40–49	50–200	>200	
Boron ppm (B)		18–24	25–75	>75	
Molybdenum ppm (Mo)					

Euonymus alatus
(Euonymus)

Plant part Most recently mature leaf — Ref: 9

Nutrient[1]	Deficient	Low	Normal	High	Excess
Nitrogen % (N)			2.40		
Phosphorus % (P)			0.21		
Potassium % (K)			1.24		
Calcium % (Ca)			1.65		
Magnesium % (Mg)			0.10		
Sulphur % (S)					
Sodium % (Na)					
Chloride % (Cl)					
Copper ppm (Cu)			12		
Zinc ppm (Zn)			34		
Manganese ppm (Mn)			73		
Iron* ppm (Fe)			304		
Boron ppm (B)			28		
Molybdenum ppm (Mo)			0.69		

Euphorbia milii
(Crown of thorns)

Plant part Most recently mature leaf — Ref: 9

Nutrient[1]	Deficient	Low	Normal	High	Excess
Nitrogen % (N)		1.70–1.99	2.00–4.00	>4.0	
Phosphorus % (P)		0.21–0.24	0.25–1.0	>1.0	
Potassium % (K)		1.00–1.49	1.50–4.0	>4.0	
Calcium % (Ca)		0.70–0.99	1.00–2.5	>2.5	
Magnesium % (Mg)		0.21–0.24	0.25–1.0	>1.0	
Sulphur % (S)		0.13–0.19	0.20–0.4	>0.4	
Sodium % (Na)					
Chloride % (Cl)					
Copper ppm (Cu)		6–9	10–50	>50	
Zinc ppm (Zn)		15–19	20–200	>200	
Manganese ppm (Mn)		20–24	25–200	>200	
Iron* ppm (Fe)		40–49	50–200	>200	
Boron ppm (B)		21–24	25–100	>100	
Molybdenum ppm (Mo)					

Euphorbia pulcherrima
(Poinsettia)

Plant part Most recently mature leaf — Ref: 7, 9

Nutrient[1]	Deficient	Low	Normal	High	Excess
Nitrogen % (N)		3.10–3.99	4.0–6.0	6.1–7.0	
Phosphorus % (P)		0.21–0.29	0.3–0.5	0.6–0.7	
Potassium % (K)		1.10–1.49	1.5–3.5	3.6–4.0	
Calcium % (Ca)		0.51–0.69	0.7–2.0	>2.0	
Magnesium % (Mg)		0.21–0.29	0.3–1.0	>1.0	
Sulphur % (S)		0.20–0.24	0.25–0.7	>0.7	
Sodium % (Na)					
Chloride % (Cl)					
Copper ppm (Cu)		1–2	3–25	>25	
Zinc ppm (Zn)		16–24	25–100	>100	
Manganese ppm (Mn)		31–44	45–300	>300	
Iron* ppm (Fe)		50–99	100–300	>300	
Boron ppm (B)	<19	20–29	30–100	100–200	>351
Molybdenum ppm (Mo)		0.6–0.9	1.0–5.0	>5.0	

Exacum
(Persian violet)

Plant part Shoot tips with one pair mature leaves and all immature leaves
Ref: 6

Nutrient[1]	Deficient	Low	Normal	High	Excess
Nitrogen % (N)			3.8–5.3		
Phosphorus % (P)			0.3–0.7		
Potassium % (K)			2.3–3.4		
Calcium % (Ca)			0.5–0.8		
Magnesium % (Mg)			0.4–0.7		
Sulphur % (S)					
Sodium % (Na)					
Chloride % (Cl)					
Copper ppm (Cu)			5–75		
Zinc ppm (Zn)			25–85		
Manganese ppm (Mn)			70–165		
Iron* ppm (Fe)			55–155		
Boron ppm (B)			25–60		
Molybdenum ppm (Mo)					

Fagus silvatica
(Beech tree)

Sampling time Current year's terminals at mid season
Plant part Most recently mature leaf
Ref: 2

Nutrient[1]	Deficient	Low	Normal	High	Excess
Nitrogen % (N)			1.90–2.50		
Phosphorus % (P)			0.15–0.30		
Potassium % (K)			1.00–1.50		
Calcium % (Ca)			0.30–1.50		
Magnesium % (Mg)			0.15–0.30		
Sulphur % (S)					
Sodium % (Na)					
Chloride % (Cl)					
Copper ppm (Cu)			5–12		
Zinc ppm (Zn)			15–50		
Manganese ppm (Mn)			35–100		
Iron* ppm (Fe)					
Boron ppm (B)			15–40		
Molybdenum ppm (Mo)			0.05–0.20		

Ficus benjamina* or *nitida
(Weeping fig)

Plant part Most recently mature leaf — Ref: 9

Nutrient[1]	Deficient	Low	Normal	High	Excess
Nitrogen % (N)		1.40–1.79	1.80–2.5	>2.5	
Phosphorus % (P)		0.08–0.09	0.10–0.5	>0.5	
Potassium % (K)		0.70–0.99	1.00–2.0	>2.0	
Calcium % (Ca)		0.70–0.99	1.00–3.0	>3.0	
Magnesium % (Mg)		0.25–0.39	0.40–1.0	>1.0	
Sulphur % (S)		0.12–0.14	0.15–0.5	>0.5	
Sodium % (Na)					
Chloride % (Cl)					
Copper ppm (Cu)		6–7	8–25	>25	
Zinc ppm (Zn)		11–14	15–200	>200	
Manganese ppm (Mn)		20–24	25–200	>200	
Iron* ppm (Fe)		20–29	30–200	>200	
Boron ppm (B)		20–29	30–75	>75	
Molybdenum ppm (Mo)					

Ficus elastica
(Rubber tree)

Plant part Petioles from most recently mature leaf — Ref: 9

Nutrient[1]	Deficient	Low	Normal	High	Excess
Nitrogen % (N)		1.00–1.29	1.30–2.5	>2.5	
Phosphorus % (P)		0.08–0.09	0.10–0.5	>0.5	
Potassium % (K)		0.40–0.59	0.60–2.1	>2.1	
Calcium % (Ca)		0.20–0.29	0.30–2.1	>2.1	
Magnesium % (Mg)		0.15–0.19	0.20–0.5	>0.5	
Sulphur % (S)		0.10–0.14	0.15–0.5	>0.5	
Sodium % (Na)					
Chloride % (Cl)					
Copper ppm (Cu)		6–7	8–25	>25	
Zinc ppm (Zn)		11–14	15–200	>200	
Manganese ppm (Mn)		15–19	20–200	>200	
Iron* ppm (Fe)		20–29	30–200	>200	
Boron ppm (B)		15–19	20–50	>50	
Molybdenum ppm (Mo)					

Ficus lyrata
(Ficus or fiddleleaf fig)

Plant part Petioles from most recently mature leaf — Ref: 9

Nutrient[1]	Deficient	Low	Normal	High	Excess
Nitrogen % (N)		1.00–1.29	1.30–2.3	>2.3	
Phosphorus % (P)		0.08–0.09	0.10–0.5	>0.5	
Potassium % (K)		0.40–0.59	0.60–2.1	2.2–3.5	
Calcium % (Ca)		0.20–0.29	0.30–1.2	>1.2	
Magnesium % (Mg)		0.15–0.19	0.2–0.5	>0.5	
Sulphur % (S)		0.10–0.14	0.15–0.5	>0.5	
Sodium % (Na)					
Chloride % (Cl)					
Copper ppm (Cu)		6–7	8–25	>25	
Zinc ppm (Zn)		11–14	15–200	>200	
Manganese ppm (Mn)		15–19	20–200	>200	
Iron* ppm (Fe)		20–29	30–200	>200	
Boron ppm (B)		15–19	20–50	>50	
Molybdenum ppm (Mo)					

Forsythia intermedia
(Forsythia)

Plant part Most recently mature leaf — Ref: 9

Nutrient[1]	Deficient	Low	Normal	High	Excess
Nitrogen % (N)			2.16		
Phosphorus % (P)			0.27		
Potassium % (K)			1.40		
Calcium % (Ca)			0.87		
Magnesium % (Mg)			0.26		
Sulphur % (S)					
Sodium % (Na)					
Chloride % (Cl)					
Copper ppm (Cu)			22		
Zinc ppm (Zn)			36		
Manganese ppm (Mn)			109		
Iron* ppm (Fe)			102		
Boron ppm (B)			14		
Molybdenum ppm (Mo)			1.18		

Fraxinus excelsior
(Ash tree)

Sampling time Current year's terminals at mid season
Plant part Most recently mature leaf

Ref: 2

Nutrient[1]	Deficient	Low	Normal	High	Excess
Nitrogen % (N)			1.70–2.20		
Phosphorus % (P)			0.15–0.30		
Potassium % (K)			1.10–1.50		
Calcium % (Ca)			0.30–1.50		
Magnesium % (Mg)			0.20–0.40		
Sulphur % (S)					
Sodium % (Na)					
Chloride % (Cl)					
Copper ppm (Cu)			6–12		
Zinc ppm (Zn)			15–50		
Manganese ppm (Mn)			30–100		
Iron* ppm (Fe)					
Boron ppm (B)			15–40		
Molybdenum ppm (Mo)			0.05–0.20		

Freesia

Plant part Most recently mature leaf

Ref: 6

Nutrient[1]	Deficient	Low	Normal	High	Excess
Nitrogen % (N)			2.7–5.6		
Phosphorus % (P)			0.4–1.2		
Potassium % (K)			3.1–5.9		
Calcium % (Ca)			0.4–1.0		
Magnesium % (Mg)			0.3–1.8		
Sulphur % (S)					
Sodium % (Na)					
Chloride % (Cl)					
Copper ppm (Cu)			5–130		
Zinc ppm (Zn)			40–110		
Manganese ppm (Mn)			30–540		
Iron* ppm (Fe)			80–115		
Boron ppm (B)			30–100		
Molybdenum ppm (Mo)					

Fuchsia

Plant part Most recently mature leaf — Ref: 6

Nutrient[1]	Deficient	Low	Normal	High	Excess
Nitrogen % (N)			2.8–4.6		
Phosphorus % (P)			0.4–0.6		
Potassium % (K)			2.2–2.5		
Calcium % (Ca)			1.6–2.4		
Magnesium % (Mg)			0.4–0.7		
Sulphur % (S)					
Sodium % (Na)					
Chloride % (Cl)					
Copper ppm (Cu)			5–10		
Zinc ppm (Zn)			30–45		
Manganese ppm (Mn)			75–220		
Iron* ppm (Fe)			95–335		
Boron ppm (B)			25–35		
Molybdenum ppm (Mo)					

Gardenia jasminoides
(Gardenia)

Plant part Most recently mature leaf — Ref: 9

Nutrient[1]	Deficient	Low	Normal	High	Excess
Nitrogen % (N)		1.20–1.49	1.50–3.0	>3.0	
Phosphorus % (P)		0.12–0.15	0.16–0.4	>0.4	
Potassium % (K)		0.80–0.99	1.00–3.0	>3.0	
Calcium % (Ca)		0.31–0.49	0.50–1.3	>1.3	
Magnesium % (Mg)		0.20–0.24	0.25–1.0	>1.0	
Sulphur % (S)		0.13–0.19	0.20–0.4	>0.4	
Sodium % (Na)					
Chloride % (Cl)					
Copper ppm (Cu)		4–5	6–40	>40	
Zinc ppm (Zn)		16–19	20–150	>150	
Manganese ppm (Mn)		40–49	50–250	>250	
Iron* ppm (Fe)		40–59	60–250	>250	
Boron ppm (B)		20–24	25–70	>70	
Molybdenum ppm (Mo)					

Gerbera jamesonii
(Gerbera or Transvaal daisy)

Plant part Most recently mature leaf — Ref: 9, 10+

Nutrient[1]	Deficient	Low	Normal	High	Excess
Nitrogen % (N)		1.20–1.49	2.7–3.1+ 1.50–3.5	>3.5	
Phosphorus % (P)		0.15–0.19	0.19–0.35+ 0.20–0.5	>0.5	
Potassium % (K)		2.00–2.49	3.06–3.64+ 2.50–4.5	>4.5	
Calcium % (Ca)		0.70–0.90	1.66–2.18+ 1.00–3.5	>3.5	
Magnesium % (Mg)		0.15–0.19	0.30–0.48+ 0.20–0.7	>0.7	
Sulphur % (S)		0.16–0.24	0.25–0.7	>0.7	
Sodium % (Na)					
Chloride % (Cl)					
Copper ppm (Cu)		4–5	6–50	>50	
Zinc ppm (Zn)		20–24	25–200	>200	
Manganese ppm (Mn)		30–39	40–250	>250	
Iron* ppm (Fe)		40–49	50–200	>200	
Boron ppm (B)		15–19	20–60	>60	
Molybdenum ppm (Mo)		<0.2	0.2–0.6	>0.6	

Gladiolus hortulanus
(Gladiolus)

Plant part Most recently mature leaf — Ref: 7, 9, 15

Nutrient[1]	Deficient	Low	Normal	High	Excess
Nitrogen % (N)		2.50–2.99	3.00–5.5	>5.5	
Phosphorus % (P)		0.23–0.24	0.25–1.0	>1.0	
Potassium % (K)		2.00–2.49	2.50–4.0	>4.0	
Calcium % (Ca)		0.40–0.49	0.50–4.5	>1.5	
Magnesium % (Mg)		0.12–0.14	0.15–0.3	>0.4	
Sulphur % (S)					
Sodium % (Na)				>0.5	
Chloride % (Cl)					>2.5
Copper ppm (Cu)		5–7	8–20	>20	
Zinc ppm (Zn)		11–19	20–200	>200	
Manganese ppm (Mn)		25–49	50–200	>200	
Iron* ppm (Fe)		40–49	50–200	>200	
Boron ppm (B)	<20	22–24	25–100	100–200	>200
Molybdenum ppm (Mo)					

Gloxinia

Plant part Most recently mature leaf — Ref: 6

Nutrient[1]	Deficient	Low	Normal	High	Excess
Nitrogen % (N)			3.3–3.8		
Phosphorus % (P)			0.3–0.5		
Potassium % (K)			4.5–5.0		
Calcium % (Ca)			1.5–2.2		
Magnesium % (Mg)			0.4–0.5		
Sulphur % (S)					
Sodium % (Na)					
Chloride % (Cl)					
Copper ppm (Cu)			5–20		
Zinc ppm (Zn)			20–35		
Manganese ppm (Mn)			95–170		
Iron* ppm (Fe)			70–150		
Boron ppm (B)			30–35		
Molybdenum ppm (Mo)					

Grevillea
(cv. Poorinda Firebird)

Plant part Most recently mature leaf — Ref: 16

Nutrient[1]	Deficient	Low	Normal	High	Excess
Nitrogen % (N)			1.76		
Phosphorus % (P)	<0.08		0.09		>0.27
Potassium % (K)			1.21		
Calcium % (Ca)			0.81		
Magnesium % (Mg)			0.23		
Sulphur % (S)					
Sodium % (Na)					
Chloride % (Cl)					
Copper ppm (Cu)					
Zinc ppm (Zn)					
Manganese ppm (Mn)					
Iron* ppm (Fe)					
Boron ppm (B)					
Molybdenum ppm (Mo)					

Gypsophila paniculata
(Baby's breath)

Plant part Most recent fully developed leaves — Ref: 9

Nutrient[1]	Deficient	Low	Normal	High	Excess
Nitrogen % (N)		3.50–4.29	4.30–6.0	>6.0	
Phosphorus % (P)		0.25–0.29	0.30–0.7	>0.7	
Potassium % (K)		2.80–3.49	3.50–4.5	>4.5	
Calcium % (Ca)		1.50–2.59	2.60–4.0	>4.0	
Magnesium % (Mg)		0.30–0.39	0.40–1.0	>1.0	
Sulphur % (S)		0.20–0.24	0.25–0.7	>0.7	
Sodium % (Na)					
Chloride % (Cl)					
Copper ppm (Cu)		6–8	9–25	>25	
Zinc ppm (Zn)		20–24	25–200	>200	
Manganese ppm (Mn)		40–49	50–200	>200	
Iron* ppm (Fe)		40–49	50–200	>200	
Boron ppm (B)		22–24	25–100	>100	
Molybdenum ppm (Mo)					

Hedera helix
(English ivy)

Plant part Most recently mature leaf — Ref: 9

Nutrient[1]	Deficient	Low	Normal	High	Excess
Nitrogen % (N)			2.64		
Phosphorus % (P)			0.33		
Potassium % (K)			2.49		
Calcium % (Ca)			0.48		
Magnesium % (Mg)			0.15		
Sulphur % (S)					
Sodium % (Na)					
Chloride % (Cl)					
Copper ppm (Cu)			26		
Zinc ppm (Zn)			50		
Manganese ppm (Mn)			59		
Iron* ppm (Fe)			389		
Boron ppm (B)			26		
Molybdenum ppm (Mo)			0.90		

***Hemerocallis* sp.**
(Day lily)

Plant part Most recently mature leaf — Ref: 9

Nutrient[1]	Deficient	Low	Normal	High	Excess
Nitrogen % (N)		2.50–2.99	3.00–5.0	>5.0	
Phosphorus % (P)		0.22–0.24	0.25–0.5	>5.0	
Potassium % (K)		2.00–2.49	2.50–4.0	>4.0	
Calcium % (Ca)		0.60–0.99	1.00–1.5	>1.5	
Magnesium % (Mg)		0.18–0.19	0.20–0.5	>0.5	
Sulphur % (S)		0.21–0.24	0.25–0.7	>0.7	
Sodium % (Na)					
Chloride % (Cl)					
Copper ppm (Cu)		5–7	8–25	>25	
Zinc ppm (Zn)		18–19	20–200	>200	
Manganese ppm (Mn)		15–24	25–300	>300	
Iron* ppm (Fe)		36–49	50–300	>300	
Boron ppm (B)		20–24	25–75	>75	
Molybdenum ppm (Mo)					

Hibiscus rosa-sinensis
(Hibiscus)

Plant part Most recently mature leaf — Ref: 7, 9

Nutrient[1]	Deficient	Low	Normal	High	Excess
Nitrogen % (N)		2.00–2.49	2.50–4.5	>4.5	
Phosphorus % (P)		0.20–0.24	0.25–1.0	>1.0	
Potassium % (K)		1.20–1.49	1.50–3.0	>3.0	
Calcium % (Ca)		0.80–0.99	1.00–3.0	>3.0	
Magnesium % (Mg)		0.21–0.24	0.25–0.8	>0.8	
Sulphur % (S)		0.13–0.19	0.20–0.5	>0.5	
Sodium % (Na)					
Chloride % (Cl)					
Copper ppm (Cu)		4–5	6–50	>50	
Zinc ppm (Zn)		16–19	20–200	>200	
Manganese ppm (Mn)		25–39	40–200	>200	
Iron* ppm (Fe)		40–49	50–200	>200	
Boron ppm (B)	<19	20–24	25–100	>100	>151
Molybdenum ppm (Mo)					

Howea forsteriana
(Kentia palm)

Plant part Middle leaflets from most recently mature frond minus petiole

Ref: 3

Nutrient[1]	Deficient	Low	Normal	High	Excess
Nitrogen % (N)	<0.84	0.85–1.19	1.20–2.75	2.76–4.00	>4.01
Phosphorus % (P)	<0.10	0.11–0.14	0.15–0.75	0.76–1.25	>1.26
Potassium % (K)	<0.59	0.60–0.84	0.85–2.25	2.26–4.00	>4.01
Calcium % (Ca)	<0.25	0.26–0.39	0.40–1.50	1.51–2.50	>2.51
Magnesium % (Mg)	<0.19	0.20–0.24	0.25–1.00	1.01–1.25	>1.26
Sulphur % (S)	<0.10	0.11–0.14	0.15–0.75	0.76–1.25	>1.26
Sodium % (Na)			0–0.20	0.21–0.50	>0.51
Chloride % (Cl)					
Copper ppm (Cu)	<4	5–7	8–200	201–500	>501
Zinc ppm (Zn)	<17	18–24	25–200	201–1000	>1001
Manganese ppm (Mn)	<39	40–49	50–250	251–1000	>1001
Iron* ppm (Fe)	<39	40–49	50–250	251–1000	>1001
Boron ppm (B)	<15	16–20	21–75	76–100	>101
Molybdenum ppm (Mo)					

Humulus lupulus
(Hop plant)

Sampling time Mid season
Plant part Most recently mature leaf

Ref: 2

Nutrient[1]	Deficient	Low	Normal	High	Excess
Nitrogen % (N)			2.50–3.50		
Phosphorus % (P)			0.35–0.60		
Potassium % (K)			2.80–3.50		
Calcium % (Ca)			1.00–2.50		
Magnesium % (Mg)			0.30–0.60		
Sulphur % (S)					
Sodium % (Na)					
Chloride % (Cl)					
Copper ppm (Cu)			6–12		
Zinc ppm (Zn)			35–80		
Manganese ppm (Mn)			30–100		
Iron* ppm (Fe)					
Boron ppm (B)			25–70		
Molybdenum ppm (Mo)			0.20–0.50		

Hydrangea macrophylla
(Hydrangea)

Plant part Most recently mature leaf — Ref: 9

Nutrient[1]	Deficient	Low	Normal	High	Excess
Nitrogen % (N)		2.5–2.99	3.00–5.5	>5.5	
Phosphorus % (P)		0.18–0.24	0.25–0.7	>0.7	
Potassium % (K)		1.50–2.10	2.20–5.0	>5.0	
Calcium % (Ca)		0.40–0.59	0.60–1.8	>1.8	
Magnesium % (Mg)		0.18–0.21	0.22–0.5	>0.5	
Sulphur % (S)		0.13–0.19	0.20–0.7	>0.7	
Sodium % (Na)					
Chloride % (Cl)					
Copper ppm (Cu)		4–5	6–50	>50	
Zinc ppm (Zn)		15–19	20–200	>200	
Manganese ppm (Mn)		40–49	50–300	>300	
Iron* ppm (Fe)		40–49	50–300	>300	
Boron ppm (B)		16–19	20–50	>50	
Molybdenum ppm (Mo)					

Ilex opaca
(American holly)

Plant part Most recently mature leaf — Ref: 9

Nutrient[1]	Deficient	Low	Normal	High	Excess
Nitrogen % (N)		<1.5	1.5–2.2	>2.2	
Phosphorus % (P)		<0.1	0.1–0.2	>0.2	
Potassium % (K)		<1.4	1.4–2.8	>2.8	
Calcium % (Ca)		<0.5	0.5–1.0	>1.0	
Magnesium % (Mg)		<0.2	0.2–0.6	>0.6	
Sulphur % (S)					
Sodium % (Na)					
Chloride % (Cl)					
Copper ppm (Cu)		<10	10–40	>40	
Zinc ppm (Zn)					
Manganese ppm (Mn)		<50	50–125	>125	
Iron* ppm (Fe)		<100	100–250	>250	
Boron ppm (B)		<35	35–65	>65	
Molybdenum ppm (Mo)					

Impatiens
(Common)

Plant part Most recently mature leaf — Ref: 6

Nutrient[1]	Deficient	Low	Normal	High	Excess
Nitrogen % (N)			4.3–5.3		
Phosphorus % (P)			0.6–0.8		
Potassium % (K)			1.8–2.8		
Calcium % (Ca)			2.9–3.3		
Magnesium % (Mg)			0.6–0.8		
Sulphur % (S)					
Sodium % (Na)					
Chloride % (Cl)					
Copper ppm (Cu)			10–15		
Zinc ppm (Zn)			65–70		
Manganese ppm (Mn)			205–490		
Iron* ppm (Fe)			405–685		
Boron ppm (B)			45–95		
Molybdenum ppm (Mo)					

Impatiens
(New Guinea)

Plant part Most recently mature leaf — Ref: 6

Nutrient[1]	Deficient	Low	Normal	High	Excess
Nitrogen % (N)			3.3–4.9		
Phosphorus % (P)			0.3–0.8		
Potassium % (K)			1.9–2.7		
Calcium % (Ca)			1.9–2.7		
Magnesium % (Mg)			0.3–0.8		
Sulphur % (S)					
Sodium % (Na)					
Chloride % (Cl)					
Copper ppm (Cu)			5–10		
Zinc ppm (Zn)			40–85		
Manganese ppm (Mn)			140–245		
Iron* ppm (Fe)			160–890		
Boron ppm (B)			50–60		
Molybdenum ppm (Mo)					

Isopogon anemonifolius

Plant part Hardened youngest fully expanded leaf — Ref: 17

Nutrient[1]	Deficient	Low	Normal	High	Excess
Nitrogen % (N)		<0.70	0.70–1.11		
Phosphorus % (P)			0.10–0.15	>0.25	
Potassium % (K)			0.97–1.11		
Calcium % (Ca)			0.35–0.43		
Magnesium % (Mg)			0.13–0.16		
Sulphur % (S)			0.19–0.23		
Sodium % (Na)			0.17–0.23		
Chloride % (Cl)			0.32–0.43		
Copper ppm (Cu)			2–5		
Zinc ppm (Zn)			25–40		
Manganese ppm (Mn)			185–380		
Iron* ppm (Fe)			27–32		
Boron ppm (B)					
Molybdenum ppm (Mo)					

Ixora coccinea
(Ixora)

Plant part Most recently mature leaf — Ref: 7, 9

Nutrient[1]	Deficient	Low	Normal	High	Excess
Nitrogen % (N)		1.30–1.79	1.80–3.0	>3.0	
Phosphorus % (P)		0.10–0.14	0.15–1.0	>1.0	
Potassium % (K)		0.80–0.99	1.00–2.5	>2.5	
Calcium % (Ca)		0.50–0.79	0.80–2.0	>2.0	
Magnesium % (Mg)		0.15–0.19	0.20–1.0	>1.0	
Sulphur % (S)		0.13–0.19	0.20–0.4	>0.4	
Sodium % (Na)					
Chloride % (Cl)					
Copper ppm (Cu)		6–9	10–50	>50	
Zinc ppm (Zn)		15–19	20–200	>200	
Manganese ppm (Mn)		40–49	50–200	>200	
Iron* ppm (Fe)		55–64	65–250	>250	
Boron ppm (B)	<19	20–24	25–100	>100	>126
Molybdenum ppm (Mo)					

Jasminum simplicifolium
(Wax jasmine)

Plant part Most recently mature leaf — Ref: 9

Nutrient[1]	Deficient	Low	Normal	High	Excess
Nitrogen % (N)		1.70–1.99	2.00–4.0	>4.0	
Phosphorus % (P)		0.15–0.17	0.18–0.5	>0.5	
Potassium % (K)		1.00–1.29	1.30–2.5	>2.5	
Calcium % (Ca)		0.40–0.69	0.70–1.5	>1.5	
Magnesium % (Mg)		0.21–0.24	0.25–1.0	>1.0	
Sulphur % (S)		0.13–0.19	0.20–0.4	>0.40	
Sodium % (Na)					
Chloride % (Cl)					
Copper ppm (Cu)		6–9	10–50	>50	
Zinc ppm (Zn)		18–19	20–200	>200	
Manganese ppm (Mn)		25–39	40–200	>200	
Iron* ppm (Fe)		40–49	50–200	>200	
Boron ppm (B)		20–24	25–75	>75	
Molybdenum ppm (Mo)					

***Juniperus* sp.**
(Juniper)

Plant part Most recently mature shoot growth — Ref: 9

Nutrient[1]	Deficient	Low	Normal	High	Excess
Nitrogen % (N)		1.10–1.49	1.50–2.50	>2.50	
Phosphorus % (P)		0.13–0.19	0.20–0.75	>0.75	
Potassium % (K)		0.70–0.99	1.00–2.50	>2.50	
Calcium % (Ca)		0.50–0.79	0.80–1.50	<1.50	
Magnesium % (Mg)		0.20–0.24	0.25–0.80	<0.80	
Sulphur % (S)		0.13–0.19	0.20–0.45	>0.45	
Sodium % (Na)					
Chloride % (Cl)					
Copper ppm (Cu)		5–7	8–50	>50	
Zinc ppm (Zn)		16–19	20–200	>200	
Manganese ppm (Mn)		20–24	25–200	>200	
Iron* ppm (Fe)		30–39	40–200	>200	
Boron ppm (B)		16–19	20–60	>60	
Molybdenum ppm (Mo)					

Kalanchoe blossfeldiana
(Kalanchoe)

Plant part 4th leaf from tip — Ref: 9

Nutrient[1]	Deficient	Low	Normal	High	Excess
Nitrogen % (N)		<2.20	2.20–4.5	>4.5	
Phosphorus % (P)		<0.25	0.25–1.0	>1.0	
Potassium % (K)		<3.00	3.00–4.0	>4.0	
Calcium % (Ca)		<2.00	2.00–4.0	>4.0	
Magnesium % (Mg)		<0.60	0.60–1.5	>1.5	
Sulphur % (S)					
Sodium % (Na)					
Chloride % (Cl)					
Copper ppm (Cu)		<4	4–10	>10	
Zinc ppm (Zn)		<30	30–100	>100	
Manganese ppm (Mn)		<70	70–100	>100	
Iron* ppm (Fe)		<55	55–100	>100	
Boron ppm (B)		<6	6–10	>10	
Molybdenum ppm (Mo)					

Larix decidua
(Larch tree)

Plant part Last season needles on uppermost lateral shoot — Ref: 2

Nutrient[1]	Deficient	Low	Normal	High	Excess
Nitrogen % (N)			1.60–2.30		
Phosphorus % (P)			0.15–0.30		
Potassium % (K)			0.50–1.10		
Calcium % (Ca)			0.60–0.90		
Magnesium % (Mg)			0.12–0.30		
Sulphur % (S)					
Sodium % (Na)					
Chloride % (Cl)					
Copper ppm (Cu)			4–10		
Zinc ppm (Zn)			20–80		
Manganese ppm (Mn)			35–200		
Iron* ppm (Fe)					
Boron ppm (B)			15–50		
Molybdenum ppm (Mo)			0.05–0.20		

Leea coccinea
(Leea or Hawaiian holly)

Plant part Most recently mature leaf — Ref: 9

Nutrient[1]	Deficient	Low	Normal	High	Excess
Nitrogen % (N)		1.05–2.19	2.20–3.3	>3.3	
Phosphorus % (P)		0.15–0.18	0.19–0.5	>0.5	
Potassium % (K)		1.20–1.49	1.50–2.8	>2.8	
Calcium % (Ca)		0.90–1.19	1.20–2.0	>2.0	
Magnesium % (Mg)		0.20–0.24	0.25–0.8	>0.8	
Sulphur % (S)		0.15–0.19	0.20–0.5	>0.5	
Sodium % (Na)					
Chloride % (Cl)					
Copper ppm (Cu)		6–9	10–50	>50	
Zinc ppm (Zn)		15–19	20–200	>200	
Manganese ppm (Mn)		20–29	30–200	>200	
Iron* ppm (Fe)		25–29	30–300	>300	
Boron ppm (B)		10–14	15–50	>50	
Molybdenum ppm (Mo)					

Leucodendron
(cv. Sundance)

Plant part Hardened youngest fully expanded leaf — Ref: 17

Nutrient[1]	Deficient	Low	Normal	High	Excess
Nitrogen % (N)		<1.00	1.00–2.20		
Phosphorus % (P)			0.05–0.13	>0.15	
Potassium % (K)			0.60–1.00		
Calcium % (Ca)			0.70–0.75		
Magnesium % (Mg)			0.20		
Sulphur % (S)			0.20		
Sodium % (Na)			0.20–0.35		
Chloride % (Cl)			0.55		
Copper ppm (Cu)			<10		
Zinc ppm (Zn)			30–40		
Manganese ppm (Mn)			350–380		
Iron* ppm (Fe)			30–35		
Boron ppm (B)					
Molybdenum ppm (Mo)					

Liatris spicata

Plant part Most recently mature leaf
Growth stage Flower buds visible — Ref: 1

Nutrient[1]	Deficient	Low	Normal	High	Excess
Nitrogen % (N)			2.7–3.3		
Phosphorus % (P)			0.19–0.20		
Potassium % (K)			1.16–2.31		
Calcium % (Ca)			1.12–1.49		
Magnesium % (Mg)				0.41–0.45	
Sulphur % (S)					
Sodium % (Na)					
Chloride % (Cl)					
Copper ppm (Cu)					
Zinc ppm (Zn)			86–94		
Manganese ppm (Mn)			163–178		
Iron* ppm (Fe)			207		
Boron ppm (B)			24–31		
Molybdenum ppm (Mo)					

Ligustrum

Plant part Most recently mature leaf — Ref: 7, 9

Nutrient[1]	Deficient	Low	Normal	High	Excess
Nitrogen % (N)		1.90–2.19	2.20–3.0	>3.0	
Phosphorus % (P)		0.16–0.19	0.20–0.5	>0.5	
Potassium % (K)		1.30–1.59	1.60–3.5	>3.5	
Calcium % (Ca)		0.50–0.69	0.70–1.5	>1.5	
Magnesium % (Mg)		0.12–0.14	0.15–0.3	>0.3	
Sulphur % (S)		0.13–0.19	0.20–0.4	>0.4	
Sodium % (Na)					
Chloride % (Cl)					
Copper ppm (Cu)		3–4	5–60	>60	
Zinc ppm (Zn)		16–19	20–200	>200	
Manganese ppm (Mn)		20–29	30–250	>250	
Iron* ppm (Fe)		40–49	50–200	>200	
Boron ppm (B)	<15	16–19	20–60	60–100	>101
Molybdenum ppm (Mo)					

Lilium longiflorum
(Easter lily)

Plant part Most recently mature leaf — Ref: 9

Nutrient[1]	Deficient	Low	Normal	High	Excess
Nitrogen % (N)		2.80–3.29	3.30–4.8	>4.8	
Phosphorus % (P)		0.15–0.24	0.25–0.7	>0.7	
Potassium % (K)		2.50–3.29	3.30–5.0	>5.0	
Calcium % (Ca)		0.35–0.59	0.60–1.5	>1.5	
Magnesium % (Mg)		0.15–0.19	0.20–0.7	>0.7	
Sulphur % (S)		0.15–0.24	0.25–0.7	>0.7	
Sodium % (Na)					
Chloride % (Cl)					
Copper ppm (Cu)		6–7	8–50	>50	
Zinc ppm (Zn)		15–19	20–200	>200	
Manganese ppm (Mn)		25–34	35–200	>200	
Iron* ppm (Fe)		50–59	60–200	>200	
Boron ppm (B)		18–24	25–75	>75	
Molybdenum ppm (Mo)					

***Limonium* sp.**
(Statice)

Plant part Most recently mature compound leaf — Ref: 9

Nutrient[1]	Deficient	Low	Normal	High	Excess
Nitrogen % (N)		3.00–3.49	3.5–6.0	>6.0	
Phosphorus % (P)		0.25–0.29	0.3–0.7	>0.7	
Potassium % (K)		2.50–2.99	3.0–5.0	>5.0	
Calcium % (Ca)		0.30–0.49	0.5–1.0	>1.0	
Magnesium % (Mg)		0.30–0.49	0.5–1.2	>1.2	
Sulphur % (S)					
Sodium % (Na)					
Chloride % (Cl)					
Copper ppm (Cu)		5–6	7–25	>25	
Zinc ppm (Zn)		20–24	25–200	>200	
Manganese ppm (Mn)		40–49	50–200	>200	
Iron* ppm (Fe)		40–49	50–200	>200	
Boron ppm (B)		16–19	20–40	>40	
Molybdenum ppm (Mo)					

Liriope muscari
(Liriope or turf lily)

Plant part Most recently mature leaf — Ref: 9

Nutrient[1]	Deficient	Low	Normal	High	Excess
Nitrogen % (N)		<2.00	2.00–3.00	>3.00	
Phosphorus % (P)		<0.25	0.25–0.35	>0.35	
Potassium % (K)		<2.0	2.00–2.90	>2.90	
Calcium % (Ca)		<0.90	0.90–1.50	>1.50	
Magnesium % (Mg)		<0.15	0.15–0.25	>0.25	
Sulphur % (S)					
Sodium % (Na)					
Chloride % (Cl)					
Copper ppm (Cu)		<6	6–15	>15	
Zinc ppm (Zn)		<19	19–35	>35	
Manganese ppm (Mn)		<25	25–75	>75	
Iron* ppm (Fe)		<50	50–200	>200	
Boron ppm (B)		<20	20–35	>35	
Molybdenum ppm (Mo)					

***Malpighia* sp.**
(Malpighia)

Plant part Most recently mature leaf — Ref: 9

Nutrient[1]	Deficient	Low	Normal	High	Excess
Nitrogen % (N)		1.50–1.99	2.00–3.5	>3.5	
Phosphorus % (P)		0.12–0.14	0.15–0.5	>0.5	
Potassium % (K)		1.20–1.49	1.50–3.0	>3.0	
Calcium % (Ca)		0.70–0.99	1.00–3.5	>3.5	
Magnesium % (Mg)		0.20–0.24	0.25–0.8	>0.8	
Sulphur % (S)		0.13–0.19	0.20–0.4	>0.4	
Sodium % (Na)					
Chloride % (Cl)					
Copper ppm (Cu)		4–5	6–50	>50	
Zinc ppm (Zn)		16–19	20–200	>200	
Manganese ppm (Mn)		15–24	25–200	>200	
Iron* ppm (Fe)		40–49	50–200	>200	
Boron ppm (B)		20–24	25–75	>75	
Molybdenum ppm (Mo)					

Mandevilla splendens
(Mandevilla or dipladenia)

Plant part Most recently mature leaf — Ref: 9

Nutrient[1]	Deficient	Low	Normal	High	Excess
Nitrogen % (N)		1.50–1.89	1.90–3.0	>3.0	
Phosphorus % (P)		0.13–0.19	0.20–0.5	>0.5	
Potassium % (K)		1.50–1.99	2.00–4.0	>4.0	
Calcium % (Ca)		0.50–0.79	0.80–1.50	>1.5	
Magnesium % (Mg)		0.20–0.24	0.25–0.5	>0.5	
Sulphur % (S)		0.13–0.19	0.20–0.4	>0.4	
Sodium % (Na)					
Chloride % (Cl)					
Copper ppm (Cu)		5–7	8–50	>50	
Zinc ppm (Zn)		16–19	20–200	>200	
Manganese ppm (Mn)		20–24	25–200	>200	
Iron* ppm (Fe)		40–49	50–200	>200	
Boron ppm (B)		20–24	25–75	>75	
Molybdenum ppm (Mo)					

Maranta leuconeura var. kerchoviana
(Rabbit's foot maranta)

Plant part Most recently mature leaf — Ref: 7, 9

Nutrient[1]	Deficient	Low	Normal	High	Excess
Nitrogen % (N)		1.50–1.99	2.0–3.0	>3.0	
Phosphorus % (P)		0.14–0.24	0.25–0.5	>0.5	
Potassium % (K)		2.20–2.99	3.00–5.5	>5.5	
Calcium % (Ca)		0.40–0.59	0.60–1.5	>1.5	
Magnesium % (Mg)		0.20–0.25	0.26–1.0	>1.0	
Sulphur % (S)		0.15–0.19	0.20–0.5	>0.5	
Sodium % (Na)					
Chloride % (Cl)					
Copper ppm (Cu)		4–7	8–50	>50	
Zinc ppm (Zn)		15–19	20–200	>200	
Manganese ppm (Mn)		40–49	50–200	>200	
Iron* ppm (Fe)		40–59	60–300	>300	
Boron ppm (B)	<19	20–24	25–50	>50	>76
Molybdenum ppm (Mo)					

Monstera deliciosa
(*Philodendron pertusum*, monstera)

Plant part Most recently mature leaf minus petiole — Ref: 9

Nutrient[1]	Deficient	Low	Normal	High	Excess
Nitrogen % (N)		2.40–2.99	3.00–5.0	>5.0	
Phosphorus % (P)		<0.20	0.20–0.4	>0.4	
Potassium % (K)		2.00–2.49	2.50–4.5	>4.5	
Calcium % (Ca)		<1.50	1.50–2.5	>2.5	
Magnesium % (Mg)		0.20–0.24	0.25–0.5	>0.5	
Sulphur % (S)		0.15–0.19	0.20–0.5	>0.5	
Sodium % (Na)					
Chloride % (Cl)					
Copper ppm (Cu)		4–6	7–50	>50	
Zinc ppm (Zn)		<25	25–200	>200	
Manganese ppm (Mn)		30–39	40–200	>200	
Iron* ppm (Fe)		45–49	50–200	>200	
Boron ppm (B)		15–16	17–60	>60	
Molybdenum ppm (Mo)					

Murraya paniculata
(Chalkas or orange jasmine)

Plant part Most recently mature leaf — Ref: 7, 9

Nutrient[1]	Deficient	Low	Normal	High	Excess
Nitrogen % (N)		1.50–1.99	2.00–3.0	>3.0	
Phosphorus % (P)		0.20–0.24	0.25–0.5	>0.5	
Potassium % (K)		0.80–0.99	1.70–3.5	>3.5	
Calcium % (Ca)		0.50–0.79	0.80–1.5	>1.5	
Magnesium % (Mg)		0.20–0.24	0.25–0.4	>0.4	
Sulphur % (S)		0.13–0.19	0.20–0.4	>0.4	
Sodium % (Na)					
Chloride % (Cl)					
Copper ppm (Cu)		3–4	7–50	>50	
Zinc ppm (Zn)		17–21	22–200	>200	
Manganese ppm (Mn)		25–49	50–250	>250	
Iron* ppm (Fe)		40–59	60–350	>350	
Boron ppm (B)	<14	17–24	25–50	>50	>101
Molybdenum ppm (Mo)					

Nephrolepis exalta var. _bostoniensis_
(Boston fern)

Plant part Most recently mature frond — Ref: 9

Nutrient[1]	Deficient	Low	Normal	High	Excess
Nitrogen % (N)		2.00–2.49	2.50–3.0	>3.0	
Phosphorus % (P)		0.15–0.24	0.25–0.7	>0.7	
Potassium % (K)		1.00–1.59	1.60–3.8	>3.8	
Calcium % (Ca)		0.50–0.79	0.80–2.5	>2.5	
Magnesium % (Mg)		0.20–0.24	0.25–1.0	>1.0	
Sulphur % (S)		0.15–0.19	0.20–0.5	>0.5	
Sodium % (Na)					
Chloride % (Cl)					
Copper ppm (Cu)		4–5	6–50	>50	
Zinc ppm (Zn)		15–19	20–200	>200	
Manganese ppm (Mn)		30–39	40–200	>200	
Iron* ppm (Fe)		40–49	50–300	>300	
Boron ppm (B)		15–19	20–70	>70	
Molybdenum ppm (Mo)					

Orchidaceae
(Cattelya orchid)

Plant part Most recently mature leaf — Ref: 9

Nutrient[1]	Deficient	Low	Normal	High	Excess
Nitrogen % (N)		1.20–1.49	1.50–2.50	>2.50	
Phosphorus % (P)		0.10–0.12	0.13–0.75	>0.75	
Potassium % (K)		1.50–1.99	2.00–3.50	>3.50	
Calcium % (Ca)		0.35–0.49	0.50–2.00	>2.00	
Magnesium % (Mg)		0.20–0.29	0.30–0.70	>0.70	
Sulphur % (S)		0.12–0.14	0.15–0.75	>0.75	
Sodium % (Na)					
Chloride % (Cl)					
Copper ppm (Cu)		2–4	5–20	>20	
Zinc ppm (Zn)		20–24	25–200	>200	
Manganese ppm (Mn)		30–39	40–200	>200	
Iron* ppm (Fe)		40–49	50–200	>200	
Boron ppm (B)		20–24	25–75	>75	
Molybdenum ppm (Mo)					

Orchidaceae
(Cymbidium orchid)

Plant part Most recently mature leaf — Ref: 9

Nutrient[1]	Deficient	Low	Normal	High	Excess
Nitrogen % (N)		1.20–1.49	1.50–2.50	>2.50	
Phosphorus % (P)		0.10–0.12	0.13–0.75	>0.75	
Potassium % (K)		1.50–1.99	2.00–3.50	>3.50	
Calcium % (Ca)		0.35–0.49	0.50–2.00	>2.00	
Magnesium % (Mg)		0.20–0.29	0.30–0.70	>0.70	
Sulphur % (S)		0.12–0.14	0.15–0.75	>0.75	
Sodium % (Na)					
Chloride % (Cl)					
Copper ppm (Cu)		2–4	5–20	>20	
Zinc ppm (Zn)		20–24	25–200	>200	
Manganese ppm (Mn)		30–39	40–200	>200	
Iron* ppm (Fe)		40–49	50–200	>200	
Boron ppm (B)		20–24	25–75	>75	
Molybdenum ppm (Mo)					

Orchidaceae
(Lady-slipper or cypripedium orchid)

Plant part Most recently mature leaf — Ref: 9

Nutrient[1]	Deficient	Low	Normal	High	Excess
Nitrogen % (N)		1.80–2.29	2.30–3.5	>3.5	
Phosphorus % (P)		0.15–0.19	0.20–0.7	>0.7	
Potassium % (K)		1.50–1.99	2.0–3.5	>3.5	
Calcium % (Ca)		0.50–0.74	0.75–0.2	>0.2	
Magnesium % (Mg)		0.15–0.19	0.20–0.7	>0.7	
Sulphur % (S)		0.15–0.19	0.20–0.7	>0.7	
Sodium % (Na)					
Chloride % (Cl)					
Copper ppm (Cu)		2–4	5–20	>20	
Zinc ppm (Zn)		20–24	25–200	>200	
Manganese ppm (Mn)		40–49	50–200	>200	
Iron* ppm (Fe)		40–49	50–200	>200	
Boron ppm (B)		20–24	25–75	>75	
Molybdenum ppm (Mo)					

Orchidaceae
(Phalenopsis or moth orchid)

Plant part Most recently mature leaf

Ref: 9

Nutrient[1]	Deficient	Low	Normal	High	Excess
Nitrogen % (N)		1.50–1.99	2.0–3.5	>3.5	
Phosphorus % (P)		0.15–0.19	0.2–0.7	>0.7	
Potassium % (K)		3.00–3.99	4.0–6.0	>6.0	
Calcium % (Ca)		1.00–1.49	1.5–2.5	>2.5	
Magnesium % (Mg)		0.30–0.39	0.4–1.0	>1.0	
Sulphur % (S)		0.15–0.19	0.2–0.7	>0.7	
Sodium % (Na)					
Chloride % (Cl)					
Copper ppm (Cu)		2–4	5–20	>20	
Zinc ppm (Zn)		15–19	20–200	>200	
Manganese ppm (Mn)		60–99	100–200	>200	
Iron* ppm (Fe)		50–74	75–200	>200	
Boron ppm (B)		20–24	25–75	>75	
Molybdenum ppm (Mo)					

Ornithogalum arabicum
(Arabian star flower)

Plant part Most recently mature leaf
Growth stage Flower buds visible

Ref: 1

Nutrient[1]	Deficient	Low	Normal	High	Excess
Nitrogen % (N)			2.0		
Phosphorus % (P)			0.25		
Potassium % (K)			3.59		
Calcium % (Ca)			2.08		
Magnesium % (Mg)			0.30		
Sulphur % (S)					
Sodium % (Na)					
Chloride % (Cl)					
Copper ppm (Cu)					
Zinc ppm (Zn)			37		
Manganese ppm (Mn)			10		
Iron* ppm (Fe)			82		
Boron ppm (B)			24		
Molybdenum ppm (Mo)					

Pelargonium hortorum
(Pelargonium)

Plant part Most recently mature leaf

Ref: 9, 14

Nutrient[1]	Deficient	Low	Normal	High	Excess
Nitrogen % (N)	<2.4	3.00–3.49	3.50–4.8	>4.8	
Phosphorus % (P)	<0.28	0.30–0.39	0.40–0.7	>0.7	
Potassium % (K)	<0.7	1.00–2.49	2.50–4.3	>4.3	
Calcium % (Ca)		0.60–0.79	0.80–1.2	>1.2	
Magnesium % (Mg)	<0.14	0.15–0.19	0.20–0.5	>0.5	
Sulphur % (S)		0.20–0.24	0.25–0.7	>0.7	
Sodium % (Na)					>0.6
Chloride % (Cl)					
Copper ppm (Cu)	<5.5	5–6	7–25	>25	
Zinc ppm (Zn)	<6	12–17	18–200	>200	
Manganese ppm (Mn)	<9	25–39	40–200	>200	>800
Iron* ppm (Fe)	<60	60–99	100–250	>250	
Boron ppm (B)	<18	18–29	30–200	201–300	>700
Molybdenum ppm (Mo)					

Pelargonium hortorum
(Pelargonium)

Sampling time Prior to or at flowering
Plant part Most recently mature leaf

Ref: 2, 7, 12

Nutrient[1]	Deficient	Low	Normal	High	Excess
Nitrogen % (N)	<0.99		2.50–3.20		
Phosphorus % (P)	<0.2		0.30–0.45		
Potassium % (K)	<0.46		1.20–2.80		
Calcium % (Ca)	<0.77		0.80–1.20		
Magnesium % (Mg)	<0.24		0.20–0.50		
Sulphur % (S)	<0.02		0.1		
Sodium % (Na)					
Chloride % (Cl)					
Copper ppm (Cu)			6–12		
Zinc ppm (Zn)			15–50		
Manganese ppm (Mn)			25–100		
Iron* ppm (Fe)	<92				
Boron ppm (B)	<17		20–50		>501
Molybdenum ppm (Mo)			0.20–0.50		

Pelargonium peltatum
(Ivy-leaved geranium)

Plant part Most recently mature leaf — Ref: 6

Nutrient[1]	Deficient	Low	Normal	High	Excess
Nitrogen % (N)			3.4–4.4		
Phosphorus % (P)			0.4–0.7		
Potassium % (K)			2.8–4.7		
Calcium % (Ca)			0.9–1.4		
Magnesium % (Mg)			0.2–0.6		
Sulphur % (S)					
Sodium % (Na)					
Chloride % (Cl)					
Copper ppm (Cu)			5–15		
Zinc ppm (Zn)			10–45		
Manganese ppm (Mn)			40–175		
Iron* ppm (Fe)			115–270		
Boron ppm (B)			30–280		
Molybdenum ppm (Mo)					

Peperomia obtusifolia
(Green or variegated peperomia)

Plant part Most recently mature leaf — Ref: 9

Nutrient[1]	Deficient	Low	Normal	High	Excess
Nitrogen % (N)		2.2–2.89	2.90–4.5	>4.5	
Phosphorus % (P)		0.2–0.24	0.25–1.0	>1.0	
Potassium % (K)		2.5–3.99	4.00–6.5	>6.5	
Calcium % (Ca)		0.7–0.99	1.00–4.0	>4.0	
Magnesium % (Mg)		0.2–0.39	0.40–1.2	>1.2	
Sulphur % (S)		0.2–0.24	0.25–0.75	>0.75	
Sodium % (Na)					
Chloride % (Cl)					
Copper ppm (Cu)		4–6	7–40	>40	
Zinc ppm (Zn)		20–24	25–200	>200	
Manganese ppm (Mn)		25–49	50–300	>300	
Iron* ppm (Fe)		25–49	50–300	>300	
Boron ppm (B)		20–24	25–50	>50	
Molybdenum ppm (Mo)					

Petunia

Sampling time Prior to or at flowering
Plant part Most recently mature leaf

Ref: 2, 6, 11

Nutrient[1]	Deficient	Low	Normal	High	Excess
Nitrogen % (N)			2.00–4.60		
Phosphorus % (P)			0.25–0.45		
Potassium % (K)			1.50–4.30		
Calcium % (Ca)			0.80–2.00		
Magnesium % (Mg)			0.20–0.50		
Sulphur % (S)					
Sodium % (Na)					
Chloride % (Cl)					
Copper ppm (Cu)			5–12		>149
Zinc ppm (Zn)			20–70		>1630
Manganese ppm (Mn)			25–100		>2560
Iron* ppm (Fe)			40–700		
Boron ppm (B)			20–50		>412
Molybdenum ppm (Mo)			0.20–0.50		

Philodendron pandurforme
(Philodendron)

Plant part Most recently mature leaf blade minus petiole

Ref: 9

Nutrient[1]	Deficient	Low	Normal	High	Excess
Nitrogen % (N)		2.30–2.49	2.50–4.5	>4.5	
Phosphorus % (P)		<0.23	0.23–0.45	>0.45	
Potassium % (K)		1.50–1.99	2.00–3.7	>3.7	
Calcium % (Ca)		<1.0	1.00–2.0	>2.0	
Magnesium % (Mg)		0.23–0.24	0.25–0.5	>0.5	
Sulphur % (S)		0.15–0.19	0.20–0.5	>0.5	
Sodium % (Na)					
Chloride % (Cl)					
Copper ppm (Cu)		5–6	7–50	>50	
Zinc ppm (Zn)		<25	25–50	>50	
Manganese ppm (Mn)		30–39	40–200	>200	
Iron* ppm (Fe)		50–59	60–200	>200	
Boron ppm (B)		15–19	20–50	>50	
Molybdenum ppm (Mo)					

Phoenix roebelenii
(Miniature date palm)

Plant part Middle leaflet from most recently mature frond, minus large petioles

Ref: 9

Nutrient[1]	Deficient	Low	Normal	High	Excess
Nitrogen % (N)		<2.00	2.00–2.8	>2.8	
Phosphorus % (P)		0.12–0.15	0.16–0.4	>0.4	
Potassium % (K)		<1.20	1.20–2.5	>2.5	
Calcium % (Ca)		0.40–0.59	0.60–1.5	>1.5	
Magnesium % (Mg)		0.11–0.17	0.18–0.3	>0.3	
Sulphur % (S)					
Sodium % (Na)					
Chloride % (Cl)					
Copper ppm (Cu)		<6	6–20	>20	
Zinc ppm (Zn)		<30	30–125	>125	
Manganese ppm (Mn)		<25	25–200	>200	
Iron* ppm (Fe)		<50	50–200	>200	
Boron ppm (B)		<16	16–30	>30	
Molybdenum ppm (Mo)					

Physostegia virginiana
(Obedient plant)

Plant part Most recently mature leaf
Growth stage Flower buds visible

Ref: 1

Nutrient[1]	Deficient	Low	Normal	High	Excess
Nitrogen % (N)			2.27		
Phosphorus % (P)			0.32		
Potassium % (K)			1.25		
Calcium % (Ca)			1.00		
Magnesium % (Mg)			0.37		
Sulphur % (S)					
Sodium % (Na)					
Chloride % (Cl)					
Copper ppm (Cu)					
Zinc ppm (Zn)			198		
Manganese ppm (Mn)			66		
Iron* ppm (Fe)			242		
Boron ppm (B)			27		
Molybdenum ppm (Mo)					

Picea abies
(Norway spruce)

Plant part One- or two-year-old needles on uppermost lateral shoot Ref: 2

Nutrient[1]	Deficient	Low	Normal	High	Excess
Nitrogen % (N)			1.35–1.70		
Phosphorus % (P)			0.13–0.25		
Potassium % (K)			0.50–1.20		
Calcium % (Ca)			0.35–0.80		
Magnesium % (Mg)			0.10–0.25		
Sulphur % (S)					
Sodium % (Na)					
Chloride % (Cl)					
Copper ppm (Cu)			4–10		
Zinc ppm (Zn)			15–60		
Manganese ppm (Mn)			50–500		
Iron* ppm (Fe)					
Boron ppm (B)			15–50		
Molybdenum ppm (Mo)			0.04–0.20		

Pilea spruceana
(cv. Angel Wings)

Plant part Most recently mature leaf Ref: 3

Nutrient[1]	Deficient	Low	Normal	High	Excess
Nitrogen % (N)			2.0–3.5		
Phosphorus % (P)			0.3–0.45		
Potassium % (K)			1.5–3.0		
Calcium % (Ca)			2.0–2.5		
Magnesium % (Mg)			1.2–1.4		
Sulphur % (S)					
Sodium % (Na)					
Chloride % (Cl)					
Copper ppm (Cu)			8–60		
Zinc ppm (Zn)			25–250		
Manganese ppm (Mn)			50–300		
Iron* ppm (Fe)			50–300		
Boron ppm (B)					
Molybdenum ppm (Mo)					

Pinus sylvestris
(Scotch or Swedish fir)

Sampling time One- or two-year-old needles
Plant part Uppermost lateral shoot

Ref: 2

Nutrient[1]	Deficient	Low	Normal	High	Excess
Nitrogen % (N)			1.40–1.70		
Phosphorus % (P)			0.14–0.30		
Potassium % (K)			0.40–0.80		
Calcium % (Ca)			0.25–0.60		
Magnesium % (Mg)			0.10–0.20		
Sulphur % (S)					
Sodium % (Na)					
Chloride % (Cl)					
Copper ppm (Cu)			4–10		
Zinc ppm (Zn)			20–70		
Manganese ppm (Mn)			50–500		
Iron* ppm (Fe)					
Boron ppm (B)			20–50		
Molybdenum ppm (Mo)			0.08–0.30		

Pittosporum tobira
(Pittosporum or mock orange)

Plant part Most recently mature leaf

Ref: 9

Nutrient[1]	Deficient	Low	Normal	High	Excess
Nitrogen % (N)		1.00–1.29	1.30–3.0	>3.5	
Phosphorus % (P)		0.20–0.24	0.25–1.0	>1.0	
Potassium % (K)		1.00–1.39	1.40–3.5	>3.5	
Calcium % (Ca)		0.50–0.79	0.80–2.5	>2.5	
Magnesium % (Mg)		0.10–0.17	0.18–1.0	>1.0	
Sulphur % (S)		0.13–0.19	0.20–0.4	>0.4	
Sodium % (Na)					
Chloride % (Cl)					
Copper ppm (Cu)		4–5	6–50	>50	
Zinc ppm (Zn)		15–19	20–200	>200	
Manganese ppm (Mn)		20–24	25–200	>200	
Iron* ppm (Fe)		25–39	40–200	>200	
Boron ppm (B)		18–19	20–75	>75	
Molybdenum ppm (Mo)					

Podocarpus macrophyllus
(Podocarpus kusamaki)

Plant part Most recently mature leaf — Ref: 9

Nutrient[1]	Deficient	Low	Normal	High	Excess
Nitrogen % (N)		1.70–1.99	2.00–3.5	>3.5	
Phosphorus % (P)		0.20–0.24	0.25–1.0	>1.0	
Potassium % (K)		0.50–0.79	0.80–2.0	>2.0	
Calcium % (Ca)		0.70–0.99	1.00–2.0	>2.0	
Magnesium % (Mg)		0.20–0.24	0.25–0.8	>0.8	
Sulphur % (S)		0.13–0.19	0.20–0.4	>0.4	
Sodium % (Na)					
Chloride % (Cl)					
Copper ppm (Cu)		5–9	10–50	>50	
Zinc ppm (Zn)		15–19	20–200	>200	
Manganese ppm (Mn)		20–24	25–200	>200	
Iron* ppm (Fe)		25–29	30–200	>200	
Boron ppm (B)		15–19	20–75	>75	
Molybdenum ppm (Mo)					

Populus
(Poplar tree)

Sampling time Current year's terminals at mid season
Plant part Most recently mature leaf — Ref: 2

Nutrient[1]	Deficient	Low	Normal	High	Excess
Nitrogen % (N)			1.80–2.50		
Phosphorus % (P)			0.18–0.30		
Potassium % (K)			1.20–1.80		
Calcium % (Ca)			0.30–1.50		
Magnesium % (Mg)			0.20–0.30		
Sulphur % (S)					
Sodium % (Na)					
Chloride % (Cl)					
Copper ppm (Cu)			6–12		
Zinc ppm (Zn)			15–50		
Manganese ppm (Mn)			35–100		
Iron* ppm (Fe)					
Boron ppm (B)			15–40		
Molybdenum ppm (Mo)			0.05–0.20		

Primula

Plant part Most recently mature leaf — Ref: 6

Nutrient[1]	Deficient	Low	Normal	High	Excess
Nitrogen % (N)			2.5–3.3		
Phosphorus % (P)			0.4–0.8		
Potassium % (K)			2.1–4.2		
Calcium % (Ca)			0.6–1.0		
Magnesium % (Mg)			0.2–0.4		
Sulphur % (S)					
Sodium % (Na)					
Chloride % (Cl)					
Copper ppm (Cu)			5–10		
Zinc ppm (Zn)			40–45		
Manganese ppm (Mn)			50–80		
Iron* ppm (Fe)			75–155		
Boron ppm (B)			30–35		
Molybdenum ppm (Mo)					

Protea
(cv. Clark's Red)

Plant part Mature leaves — Ref: 17

Nutrient[1]	Deficient	Low	Normal	High	Excess
Nitrogen % (N)		<1.24	1.24–1.93		
Phosphorus % (P)			0.09–0.14	>0.14	
Potassium % (K)			1.23–1.28		
Calcium % (Ca)			0.40–0.60		
Magnesium % (Mg)			0.10–0.11		
Sulphur % (S)			0.10–0.11		
Sodium % (Na)			0.22		
Chloride % (Cl)					
Copper ppm (Cu)			<10		
Zinc ppm (Zn)			39–45		
Manganese ppm (Mn)			180–280		
Iron* ppm (Fe)			20–22		
Boron ppm (B)					
Molybdenum ppm (Mo)					

Protea
(cv. Masquerade)

Plant part Mature leaves Ref: 17

Nutrient[1]	Deficient	Low	Normal	High	Excess
Nitrogen % (N)		<0.90	0.90–2.30		
Phosphorus % (P)			0.06–0.17	>0.17	
Potassium % (K)			0.75–1.40		
Calcium % (Ca)			0.50–0.70		
Magnesium % (Mg)					
Sulphur % (S)					
Sodium % (Na)			0.20–0.21		
Chloride % (Cl)			0.35–0.37		
Copper ppm (Cu)			<10		
Zinc ppm (Zn)			20–40		
Manganese ppm (Mn)			170–280		
Iron* ppm (Fe)			20–25		
Boron ppm (B)					
Molybdenum ppm (Mo)					

Protea neriifolia
(cv. Pink Ice)

Sampling time December–February, May–August (Australia)
Plant part Youngest fully expanded leaf Ref: 13

Nutrient[1]	Dec–Feb	May–Aug
Nitrogen % (N)	0.82–0.83	0.77–0.86
Phosphorus % (P)	0.06–0.07	0.05–0.06
Potassium % (K)	0.37–0.41	0.18–0.21
Calcium % (Ca)	0.46–0.51	0.63–0.68
Magnesium % (Mg)		
Sulphur % (S)	0.11–0.13	0.09–0.10
Sodium % (Na)		0.14–0.18
Chloride % (Cl)		
Copper ppm (Cu)		3.5–4.5
Zinc ppm (Zn)	12–15	
Manganese ppm (Mn)	43–44	
Iron* ppm (Fe)		51–54
Boron ppm (B)		
Molybdenum ppm (Mo)		

Protea neriifolia
(cv. Pink Ice, cv. Satin Mink)

Plant part Most recently mature leaf — Ref: 5, 18

Nutrient[1]	Deficient	Low	Normal	High	Excess
Nitrogen % (N)					
Phosphorus % (P)			0.03–0.06		
(Pink Ice)		<0.19	0.19–0.29	0.29–0.36	>0.36
(Satin Mink)		<0.06	0.06–0.27	0.27–0.57	>0.57
Potassium % (K)			0.3–0.7		
Calcium % (Ca)			0.5–1.0		
Magnesium % (Mg)			0.1–0.3		
Sulphur % (S)					
Sodium % (Na)					
Chloride % (Cl)					
Copper ppm (Cu)					
Zinc ppm (Zn)					
Manganese ppm (Mn)					
Iron* ppm (Fe)					
Boron ppm (B)					
Molybdenum ppm (Mo)					

Pseudotsuga taxifolia
(Douglas fir or Oregon pine)

Plant part One- to two-year-old needles on uppermost lateral shoot — Ref: 2

Nutrient[1]	Deficient	Low	Normal	High	Excess
Nitrogen % (N)			1.10–1.70		
Phosphorus % (P)			0.12–0.30		
Potassium % (K)			0.60–1.10		
Calcium % (Ca)			0.20–0.60		
Magnesium % (Mg)			0.10–0.25		
Sulphur % (S)					
Sodium % (Na)					
Chloride % (Cl)					
Copper ppm (Cu)			2–10		
Zinc ppm (Zn)			15–80		
Manganese ppm (Mn)			50–500		
Iron* ppm (Fe)					
Boron ppm (B)			20–40		
Molybdenum ppm (Mo)			0.05–0.20		

Pteris sp.
(Pteris fern)

Plant part Most recently mature frond — Ref: 9

Nutrient[1]	Deficient	Low	Normal	High	Excess
Nitrogen % (N)		1.80–2.2	2.30–3.0	>3.0	
Phosphorus % (P)		0.15–0.2	0.21–0.3	>0.3	
Potassium % (K)		<1.0	1.00–2.0	>2.0	
Calcium % (Ca)		<2.0	2.00–3.0	>3.0	
Magnesium % (Mg)		<0.25	0.25–0.4	>0.4	
Sulphur % (S)					
Sodium % (Na)					
Chloride % (Cl)					
Copper ppm (Cu)		<6	6–30	>30	
Zinc ppm (Zn)		<25	25–150	>150	
Manganese ppm (Mn)		<70	70–300	>300	
Iron* ppm (Fe)		<40	40–300	>300	
Boron ppm (B)		<20	20–30	>30	
Molybdenum ppm (Mo)					

Rhapis excelsa
(Rhapis or lady palm)

Plant part Most recently mature frond — Ref: 3

Nutrient[1]	Deficient	Low	Normal	High	Excess
Nitrogen % (N)	<0.84	0.85–1.19	1.20–2.75	2.76–4.00	>4.01
Phosphorus % (P)	<0.10	0.11–0.14	0.15–0.75	0.76–1.25	>1.26
Potassium % (K)	<0.59	0.60–0.84	0.85–2.25	2.26–4.00	>4.01
Calcium % (Ca)	<0.25	0.26–0.39	0.40–1.50	1.51–2.50	>2.51
Magnesium % (Mg)	<0.19	0.20–0.24	0.25–1.00	1.01–1.25	>1.26
Sulphur % (S)	<0.10	0.11–0.14	0.15–0.75	0.76–1.25	>1.26
Sodium % (Na)			0–0.20	0.21–0.50	>0.51
Chloride % (Cl)					
Copper ppm (Cu)	<4	5–7	8–200	201–500	>501
Zinc ppm (Zn)	<17	18–24	25–200	201–1000	>1001
Manganese ppm (Mn)	<39	40–49	50–250	251–1000	>1001
Iron* ppm (Fe)	<39	40–49	50–250	251–1000	>1001
Boron ppm (B)	<15	16–20	21–75	76–100	>101
Molybdenum ppm (Mo)					

***Rhododendron* sp.**
(Rhododendron)

Plant part Most recently mature leaf — Ref: 7, 9

Nutrient[1]	Deficient	Low	Normal	High	Excess
Nitrogen % (N)			1.60–2.10		
Phosphorus % (P)			0.21–0.29		
Potassium % (K)			0.40–0.80		
Calcium % (Ca)			0.80–1.40		
Magnesium % (Mg)			0.19–0.27		
Sulphur % (S)					
Sodium % (Na)					
Chloride % (Cl)					
Copper ppm (Cu)			1–3		
Zinc ppm (Zn)			21–37		
Manganese ppm (Mn)			602–1511		
Iron* ppm (Fe)			93–245		
Boron ppm (B)	<19		33–47		>101
Molybdenum ppm (Mo)					

Rhododendron indicum
(Azalea)

Plant part Most recently mature leaf — Ref: 9

Nutrient[1]	Deficient	Low	Normal	High	Excess
Nitrogen % (N)		1.00–1.49	1.50–2.5	>2.5	
Phosphorus % (P)		0.15–0.19	0.20–0.5	>0.5	
Potassium % (K)		0.30–0.49	0.50–1.5	>1.5	
Calcium % (Ca)		0.30–0.49	0.50–1.5	>1.5	
Magnesium % (Mg)		0.20–0.24	0.25–1.0	>1.0	
Sulphur % (S)		0.13–0.19	0.20–0.5	>0.5	
Sodium % (Na)					
Chloride % (Cl)					
Copper ppm (Cu)		4–5	6–25	>25	
Zinc ppm (Zn)		15–19	20–200	>200	
Manganese ppm (Mn)		30–39	40–200	>200	
Iron* ppm (Fe)		40–49	50–250	>250	
Boron ppm (B)		20–24	25–75	>75	
Molybdenum ppm (Mo)					

Rhododendron sp. hybrids
(Azalea, Indian hybrids)

Plant part Most recently mature leaf — Ref: 9

Nutrient[1]	Deficient	Low	Normal	High	Excess
Nitrogen % (N)		1.60–1.99	2.00–3.0	>3.0	
Phosphorus % (P)		0.20–0.29	0.30–0.5	>0.5	
Potassium % (K)		1.20–1.49	1.50–2.5	>2.5	
Calcium % (Ca)		0.30–0.69	0.70–1.5	>1.5	
Magnesium % (Mg)		0.20–0.24	0.25–0.6	>0.6	
Sulphur % (S)		0.13–0.19	0.20–0.5	>0.5	
Sodium % (Na)					
Chloride % (Cl)					
Copper ppm (Cu)		4–5	6–25	>25	
Zinc ppm (Zn)		15–19	20–250	>250	
Manganese ppm (Mn)		30–39	40–200	>200	
Iron* ppm (Fe)		40–49	50–250	>250	
Boron ppm (B)		20–24	25–50	>50	
Molybdenum ppm (Mo)					

Rosa odorata
(Rose)

Sampling time Flower bud pea size
Plant part Most recently mature compound leaf — Ref: 9

Nutrient[1]	Deficient	Low	Normal	High	Excess
Nitrogen % (N)		2.80–2.99	3.00–5.0	>5.0	
Phosphorus % (P)		0.21–0.24	0.25–0.5	>3.0	
Potassium % (K)		1.10–1.49	1.50–3.0	>3.0	
Calcium % (Ca)		0.80–0.99	1.00–2.0	>2.0	
Magnesium % (Mg)		0.21–0.24	0.25–0.5	>0.5	
Sulphur % (S)		0.20–0.24	0.25–0.7	>0.7	
Sodium % (Na)					
Chloride % (Cl)					
Copper ppm (Cu)		4–5	7–25	>25*	
Zinc ppm (Zn)		15–17	17–100	>100	
Manganese ppm (Mn)		25–29	30–200	>200	
Iron* ppm (Fe)		50–59	60–200	>200	
Boron ppm (B)	<24	25–29	30–60	>60	>126
Molybdenum ppm (Mo)					

Rosa floribunda
(Floribunda rose)

Plant part 2nd and 3rd leaves from flowering shoots — Ref: 8

Nutrient[1]	Deficient	Low	Normal	High	Excess
Nitrogen % (N)	<2.14		3.0–3.3		
Phosphorus % (P)	<0.14		0.28–0.36		
Potassium % (K)	<1.01		1.9–2.2		
Calcium % (Ca)	<0.39		1.3		
Magnesium % (Mg)	<0.2		0.3		
Sulphur % (S)					
Sodium % (Na)			0.4		
Chloride % (Cl)			0.03		
Copper ppm (Cu)					
Zinc ppm (Zn)					
Manganese ppm (Mn)					
Iron* ppm (Fe)					
Boron ppm (B)					
Molybdenum ppm (Mo)					

Rosa spp.
(Hybrid tea rose)

Plant part 2nd and 3rd leaves from flowering shoots — Ref: 8

Nutrient[1]	Deficient	Low	Normal	High	Excess
Nitrogen % (N)			3.4–3.6		
Phosphorus % (P)			0.28–0.36		
Potassium % (K)			2.3–2.6		
Calcium % (Ca)			0.95–1.3		
Magnesium % (Mg)			0.4		
Sulphur % (S)					
Sodium % (Na)					
Chloride % (Cl)			0.01		
Copper ppm (Cu)					
Zinc ppm (Zn)					
Manganese ppm (Mn)					
Iron* ppm (Fe)					
Boron ppm (B)					
Molybdenum ppm (Mo)					

Rumohra adiantiformis
(Leatherleaf fern)

Plant part Most recently mature frond — Ref: 7, 9

Nutrient[1]	Deficient	Low	Normal	High	Excess
Nitrogen % (N)		1.50–1.99	2.00–3.0	>3.0	
Phosphorus % (P)		0.15–0.24	0.25–0.5	>0.5	
Potassium % (K)		1.50–1.99	2.00–4.0	>4.0	
Calcium % (Ca)		0.30–0.49	0.50–1.0	>1.0	
Magnesium % (Mg)		0.10–0.19	0.20–0.8	>0.8	
Sulphur % (S)		0.15–0.19	0.20–0.5	>0.5	
Sodium % (Na)					
Chloride % (Cl)					
Copper ppm (Cu)		3–4	5–50	>50	
Zinc ppm (Zn)		15–19	20–200	>200	
Manganese ppm (Mn)		20–29	30–300	>300	
Iron* ppm (Fe)		15–19	20–200	>200	
Boron ppm (B)	<14	15–19	20–50	>50	>76
Molybdenum ppm (Mo)					

Saintpaulia ionantha
(African violet)

Plant part Most recently mature leaf — Ref: 9

Nutrient[1]	Deficient	Low	Normal	High	Excess
Nitrogen % (N)		2.50–2.99	3.00–6.0	>6.0	
Phosphorus % (P)		0.25–0.29	0.30–0.7	>0.7	
Potassium % (K)		2.20–2.99	3.00–6.5	>6.5	
Calcium % (Ca)		0.70–0.99	1.00–2.0	>2.0	
Magnesium % (Mg)		0.25–0.34	0.35–0.75	>0.75	
Sulphur % (S)		0.25–0.29	0.30–0.7	>0.7	
Sodium % (Na)					
Chloride % (Cl)					
Copper ppm (Cu)		5–7	8–35	>35	
Zinc ppm (Zn)		20–24	25–200	>200	
Manganese ppm (Mn)		30–39	40–200	>200	
Iron* ppm (Fe)		40–49	50–200	>200	
Boron ppm (B)		20–24	25–75	>75	
Molybdenum ppm (Mo)					

Salvia splendens

Plant part Most recently mature leaf — Ref: 9

Nutrient[1]	Deficient	Low	Normal	High	Excess
Nitrogen % (N)		2.50–2.99	3.00–4.5	>4.5	
Phosphorus % (P)		0.22–0.29	0.30–0.7	>0.7	
Potassium % (K)		3.00–3.49	3.50–5.0	>5.0	
Calcium % (Ca)		1.00–1.49	1.50–2.5	>2.5	
Magnesium % (Mg)		0.20–0.24	0.25–0.6	>0.6	
Sulphur % (S)					
Sodium % (Na)					
Chloride % (Cl)					
Copper ppm (Cu)		5–6	7–50	>50	
Zinc ppm (Zn)		20–24	25–200	>200	
Manganese ppm (Mn)		25–29	30–200	>200	
Iron* ppm (Fe)		50–59	60–300	>300	
Boron ppm (B)		20–24	25–75	>75	
Molybdenum ppm (Mo)		<2	2–4	>4	

Sansevieria laurentii
(Sansevieria)

Plant part Most recently mature leaf — Ref: 9

Nutrient[1]	Deficient	Low	Normal	High	Excess
Nitrogen % (N)		1.50–1.69	1.70–3.0	>3.0	
Phosphorus % (P)		0.10–0.14	0.15–0.4	>0.4	
Potassium % (K)		1.50–1.99	2.00–3.0	>3.0	
Calcium % (Ca)		0.70–0.99	1.00–2.0	>2.0	
Magnesium % (Mg)		0.20–0.29	0.30–0.6	>0.6	
Sulphur % (S)		0.12–0.19	0.20–0.5	>0.5	
Sodium % (Na)					
Chloride % (Cl)					
Copper ppm (Cu)		5–9	10–50	>50	
Zinc ppm (Zn)		20–24	25–200	>200	
Manganese ppm (Mn)		25–39	40–300	>300	
Iron* ppm (Fe)		30–49	50–300	>300	
Boron ppm (B)		15–19	20–50	>50	
Molybdenum ppm (Mo)					

Scabiosa caucasica
(Perennial scabious)

Plant part Most recently mature leaf
Growth stage Flower buds visible

Ref: 1

Nutrient[1]	Deficient	Low	Normal	High	Excess
Nitrogen % (N)			2.81		
Phosphorus % (P)			0.22		
Potassium % (K)			2.11		
Calcium % (Ca)			0.20		
Magnesium % (Mg)			0.36		
Sulphur % (S)					
Sodium % (Na)					
Chloride % (Cl)					
Copper ppm (Cu)					
Zinc ppm (Zn)			19		
Manganese ppm (Mn)			116		
Iron* ppm (Fe)			400		
Boron ppm (B)			27		
Molybdenum ppm (Mo)					

Schefflera actinophylla
(Schefflera or umbrella tree)

Plant part Most recently mature leaf

Ref: 7, 9

Nutrient[1]	Deficient	Low	Normal	High	Excess
Nitrogen % (N)		2.00–2.49	2.00–3.5	>3.5	
Phosphorus % (P)		0.15–0.19	0.20–0.5	>0.5	
Potassium % (K)		1.80–2.29	2.30–4.0	>4.0	
Calcium % (Ca)		0.80–0.99	1.00–1.5	>1.5	
Magnesium % (Mg)		0.20–0.24	0.25–0.75	>0.75	
Sulphur % (S)		0.15–0.20	0.21–0.8	>0.8	
Sodium % (Na)					
Chloride % (Cl)					
Copper ppm (Cu)		6–9	10–60	>60	
Zinc ppm (Zn)		15–19	20–200	>200	
Manganese ppm (Mn)		30–39	40–300	>300	
Iron* ppm (Fe)		40–49	50–300	>300	
Boron ppm (B)	<14	15–19	20–60	>60	>101
Molybdenum ppm (Mo)					

Schefflera arboricola

Plant part Most recently mature leaf — Ref: 14, 19

Nutrient[1]	Deficient	Low	Normal	High	Excess
Nitrogen % (N)	<2.0		2.8–3.7		
Phosphorus % (P)			0.26–0.35		
Potassium % (K)	<1.5		2.5–3.5		
Calcium % (Ca)			1.5–2.0		
Magnesium % (Mg)			0.4–1.0		
Sulphur % (S)					
Sodium % (Na)					
Chloride % (Cl)					
Copper ppm (Cu)			8–60		
Zinc ppm (Zn)			25–250		
Manganese ppm (Mn)			50–300		>1000
Iron* ppm (Fe)			50–300		>1900
Boron ppm (B)	<20		25–60		
Molybdenum ppm (Mo)					

Schlumbergera bridgesii
(Christmas cactus)

Plant part Most recently mature leaf — Ref: 9

Nutrient[1]	Deficient	Low	Normal	High	Excess
Nitrogen % (N)		2.50–2.70	2.8–4.5	>4.5	
Phosphorus % (P)		0.40–0.50	0.6–1.0	>1.0	
Potassium % (K)		3.20–3.90	4.0–6.0	>6.0	
Calcium % (Ca)		0.40–0.70	0.8–1.5	>1.5	
Magnesium % (Mg)		0.25–0.39	0.4–1.0	>1.0	
Sulphur % (S)		0.20–0.24	0.25–0.5	>0.5	
Sodium % (Na)					
Chloride % (Cl)					
Copper ppm (Cu)		6–9	10–30	>30	
Zinc ppm (Zn)		20–24	25–100	>100	
Manganese ppm (Mn)		40–59	60–300	>300	
Iron* ppm (Fe)		50–74	75–300	>300	
Boron ppm (B)		15–19	20–50	>50	
Molybdenum ppm (Mo)					

Scindapsus aureus
(Marble queen or pothos)

Plant part Most recently mature leaf — Ref: 9

Nutrient[1]	Deficient	Low	Normal	High	Excess
Nitrogen % (N)		2.20–2.69	2.7–3.5	>3.5	
Phosphorus % (P)		0.12–0.19	0.2–0.5	>0.5	
Potassium % (K)		2.5–2.99	3.0–4.5	>4.5	
Calcium % (Ca)		0.70–0.99	1.0–2.0	>2.0	
Magnesium % (Mg)		0.20–0.29	0.3–1.0	>1.0	
Sulphur % (S)					
Sodium % (Na)					
Chloride % (Cl)					
Copper ppm (Cu)		4–5	6–50	>50	
Zinc ppm (Zn)		15–19	20–200	>200	
Manganese ppm (Mn)		30–49	50–300	>300	
Iron* ppm (Fe)		30–49	50–300	>300	
Boron ppm (B)		15–19	20–60	>60	
Molybdenum ppm (Mo)					

Sinningia speciosa
(Gloxinia)

Plant part Most recently mature leaf — Ref: 9

Nutrient[1]	Deficient	Low	Normal	High	Excess
Nitrogen % (N)		2.50–2.99	3.00–5.0	>5.0	
Phosphorus % (P)		0.20–0.24	0.25–0.7	>0.7	
Potassium % (K)		2.10–2.49	2.50–5.0	>5.0	
Calcium % (Ca)		0.70–0.99	1.00–3.0	>3.0	
Magnesium % (Mg)		0.15–0.34	0.35–0.7	>0.7	
Sulphur % (S)		0.20–0.24	0.25–0.7	>0.7	
Sodium % (Na)					
Chloride % (Cl)					
Copper ppm (Cu)		5–7	8–25	>25	
Zinc ppm (Zn)		18–19	20–50	>50	
Manganese ppm (Mn)		25–49	50–300	>300	
Iron* ppm (Fe)		45–49	50–200	>200	
Boron ppm (B)		22–24	25–50	>50	
Molybdenum ppm (Mo)					

Solidago hybrids (Golden rod)

Plant part Most recently mature leaf
Growth stage Flower buds visible

Ref: 1

Nutrient[1]	Deficient	Low	Normal	High	Excess
Nitrogen % (N)			2.7–3.6		
Phosphorus % (P)			0.27–0.46		
Potassium % (K)			3.82–4.71		
Calcium % (Ca)			0.87–1.23		
Magnesium % (Mg)			0.30–0.43		
Sulphur % (S)					
Sodium % (Na)					
Chloride % (Cl)					
Copper ppm (Cu)					
Zinc ppm (Zn)			25–68		
Manganese ppm (Mn)			115–282		
Iron* ppm (Fe)			200–202		
Boron ppm (B)			24–30		
Molybdenum ppm (Mo)					

Spathiphyllum sp.

Plant part Most recently mature leaf
Growth stage Young plants <4 months

Ref: 9

Nutrient[1]	Deficient	Low	Normal	High	Excess
Nitrogen % (N)		3.50–3.79	3.80–5.0	>5.0	
Phosphorus % (P)		0.20–0.24	0.25–1.0	>1.0	
Potassium % (K)		3.20–3.99	4.00–6.0	>6.0	
Calcium % (Ca)		0.70–0.99	1.00–2.00	>2.0	
Magnesium % (Mg)		0.20–0.24	0.25–1.0	>1.0	
Sulphur % (S)		0.20–0.24	0.25–0.5	>0.5	
Sodium % (Na)					
Chloride % (Cl)					
Copper ppm (Cu)		5–7	8–50	>50	
Zinc ppm (Zn)		18–24	25–200	>200	
Manganese ppm (Mn)		25–39	40–300	>300	
Iron* ppm (Fe)		25–49	50–300	>300	
Boron ppm (B)		20–24	25–70	>70	
Molybdenum ppm (Mo)					

Spathiphyllum sp.

Plant part Most recently mature leaf
Growth stage >4 months

Ref: 9

Nutrient[1]	Deficient	Low	Normal	High	Excess
Nitrogen % (N)		3.00–3.29	3.3–4.5	>4.5	
Phosphorus % (P)		0.16–0.19	0.2–0.8	>0.8	
Potassium % (K)		2.00–2.29	2.3–4.0	>4.0	
Calcium % (Ca)		0.70–0.99	1.0–2.0	>2.0	
Magnesium % (Mg)		0.20–0.24	0.25–0.5	>0.5	
Sulphur % (S)		0.16–0.19	0.2–0.5	>0.5	
Sodium % (Na)					
Chloride % (Cl)					
Copper ppm (Cu)		5–7	8–50	>50	
Zinc ppm (Zn)		18–24	25–200	>200	
Manganese ppm (Mn)		25–39	40–300	>300	
Iron* ppm (Fe)		25–49	50–300	>300	
Boron ppm (B)		20–24	25–70	>70	
Molybdenum ppm (Mo)					

Spirea sp.
(Spirea)

Plant part Most recently mature leaf

Ref: 9

Nutrient[1]	Deficient	Low	Normal	High	Excess
Nitrogen % (N)			2.24–2.42		
Phosphorus % (P)			0.26–0.34		
Potassium % (K)			1.1–1.55		
Calcium % (Ca)			0.92–0.99		
Magnesium % (Mg)			0.33–0.36		
Sulphur % (S)					
Sodium % (Na)					
Chloride % (Cl)					
Copper ppm (Cu)			13–21		
Zinc ppm (Zn)			27–61		
Manganese ppm (Mn)			43–312		
Iron* ppm (Fe)			102–461		
Boron ppm (B)			25–35		
Molybdenum ppm (Mo)			2.1–3.7		

Stirlingia latifolia

Plant part Most recently mature leaf — Ref: 17

Nutrient[1]	Deficient	Low	Normal	High	Excess
Nitrogen % (N)		<0.80	0.80–1.09		
Phosphorus % (P)			0.10–0.11	>0.15	
Potassium % (K)			1.30–1.50		
Calcium % (Ca)			0.40–0.73		
Magnesium % (Mg)			0.13–0.18		
Sulphur % (S)			0.13–0.16		
Sodium % (Na)			0.28–0.36		
Chloride % (Cl)			0.33–0.39		
Copper ppm (Cu)			5–10		
Zinc ppm (Zn)			25–40		
Manganese ppm (Mn)			165–190		
Iron* ppm (Fe)			15–130		
Boron ppm (B)					
Molybdenum ppm (Mo)					

Strelitzia reginae
(Bird of paradise)

Plant part Most recently mature leaf — Ref: 9

Nutrient[1]	Deficient	Low	Normal	High	Excess
Nitrogen % (N)		0.70–0.90	1.00–2.5	>2.5	
Phosphorus % (P)		0.13–0.19	0.20–0.4	>0.4	
Potassium % (K)		1.20–1.40	1.50–3.0	>3.0	
Calcium % (Ca)		0.50–0.70	0.80–3.0	>3.0	
Magnesium % (Mg)		0.15–0.17	0.18–0.75	>0.75	
Sulphur % (S)		0.13–0.19	0.20–0.4	>0.4	
Sodium % (Na)					
Chloride % (Cl)					
Copper ppm (Cu)		5–7	8–30	>30	
Zinc ppm (Zn)		15–17	18–200	>200	
Manganese ppm (Mn)		35–44	45–200	>200	
Iron* ppm (Fe)		40–49	50–200	>200	
Boron ppm (B)		16–19	20–75	>75	
Molybdenum ppm (Mo)					

Streptocarpus sp.

Plant part Most recently mature leaf — Ref: 6

Nutrient[1]	Deficient	Low	Normal	High	Excess
Nitrogen % (N)			2.0–3.5		
Phosphorus % (P)			0.1–0.7		
Potassium % (K)			4.8–5.5		
Calcium % (Ca)			1.2–1.9		
Magnesium % (Mg)			0.3–0.5		
Sulphur % (S)					
Sodium % (Na)					
Chloride % (Cl)					
Copper ppm (Cu)			15–20		
Zinc ppm (Zn)			85–130		
Manganese ppm (Mn)			130–300		
Iron* ppm (Fe)			90–260		
Boron ppm (B)			55–65		
Molybdenum ppm (Mo)					

Stromanthe amabilis

Plant part Most recently mature leaf — Ref: 19

Nutrient[1]	Deficient	Low	Normal	High	Excess
Nitrogen % (N)			2.5–3.0		
Phosphorus % (P)			0.2–0.5		
Potassium % (K)			3.0–4.0		
Calcium % (Ca)			0.1–0.2		
Magnesium % (Mg)			0.3–0.5		
Sulphur % (S)					
Sodium % (Na)					
Chloride % (Cl)					
Copper ppm (Cu)			8–60		
Zinc ppm (Zn)			25–250		
Manganese ppm (Mn)			50–300		
Iron* ppm (Fe)			50–300		
Boron ppm (B)					
Molybdenum ppm (Mo)					

Swainsona formosa
(Sturt's desert pea)

Plant part Youngest fully mature leaf and petiole
Growth stage First flower

Ref: 20

Nutrient[1]	Deficient	Low	Normal	High	Excess
Nitrogen % (N)					
Phosphorus % (P)		<0.08	0.17–0.21	>0.5	
Potassium % (K)					
Calcium % (Ca)					
Magnesium % (Mg)					
Sulphur % (S)					
Sodium % (Na)					
Chloride % (Cl)					
Copper ppm (Cu)					
Zinc ppm (Zn)					
Manganese ppm (Mn)					
Iron* ppm (Fe)					
Boron ppm (B)					
Molybdenum ppm (Mo)					

Syngonium podophyllum
(Nepthytis)

Plant part Most recently mature leaf

Ref: 9

Nutrient[1]	Deficient	Low	Normal	High	Excess
Nitrogen % (N)		2.00–2.49	2.5–3.5	>3.5	
Phosphorus % (P)		0.15–0.19	0.2–0.5	>0.5	
Potassium % (K)		2.50–2.99	3.0–4.5	>4.5	
Calcium % (Ca)		0.30–0.39	0.4–1.5	>1.5	
Magnesium % (Mg)		0.25–0.29	0.3–0.6	>0.6	
Sulphur % (S)		0.15–0.19	0.2–0.5	>0.5	
Sodium % (Na)					
Chloride % (Cl)					
Copper ppm (Cu)		6–9	10–60	>60	
Zinc ppm (Zn)					
Manganese ppm (Mn)		40–49	50–300	>300	
Iron* ppm (Fe)		40–49	50–300	>300	
Boron ppm (B)		20–24	25–50	>50	
Molybdenum ppm (Mo)					

***Syringa* sp.**
(Lilac)

Plant part Current season leaf — Ref: 9

Nutrient[1]	Deficient	Low	Normal	High	Excess
Nitrogen % (N)		<1.6	1.6–2.5	>2.5	
Phosphorus % (P)		<0.25	0.25–0.4	>0.4	
Potassium % (K)		<1.0	1.00–1.8	>1.8	
Calcium % (Ca)		<0.6	0.60–1.2	>1.2	
Magnesium % (Mg)		<0.2	0.20–0.4	>0.4	
Sulphur % (S)					
Sodium % (Na)					
Chloride % (Cl)					
Copper ppm (Cu)		<8	8–25	>25	
Zinc ppm (Zn)		<25	25–75	>75	
Manganese ppm (Mn)		<30	35–300	>300	
Iron* ppm (Fe)		<75	75–300	>300	
Boron ppm (B)		<18	18–40	>40	
Molybdenum ppm (Mo)		<1	1–4	>4	

Taxus baccata
(Yew tree)

Sampling time One- to two-year-old needles
Plant part Uppermost lateral shoot — Ref: 2

Nutrient[1]	Deficient	Low	Normal	High	Excess
Nitrogen % (N)			1.60–2.50		
Phosphorus % (P)			0.14–0.25		
Potassium % (K)			0.90–2.00		
Calcium % (Ca)			0.25–1.00		
Magnesium % (Mg)			0.10–0.25		
Sulphur % (S)					
Sodium % (Na)					
Chloride % (Cl)					
Copper ppm (Cu)			5–12		
Zinc ppm (Zn)			25–100		
Manganese ppm (Mn)			40–500		
Iron* ppm (Fe)					
Boron ppm (B)			15–60		
Molybdenum ppm (Mo)			0.07–0.40		

Taxus media
(Taxus)

Plant part Current season leaf — Ref: 9

Nutrient[1]	Deficient	Low	Normal	High	Excess
Nitrogen % (N)		<2.0	2.00–4.0	>4.0	
Phosphorus % (P)		<0.3	0.30–0.5	>0.5	
Potassium % (K)		<1.0	1.00–2.0	>2.0	
Calcium % (Ca)		<0.6	0.60–1.0	>1.0	
Magnesium % (Mg)		<0.2	0.20–0.3	>0.3	
Sulphur % (S)					
Sodium % (Na)					
Chloride % (Cl)					
Copper ppm (Cu)		<10	10–20	>20	
Zinc ppm (Zn)		<25	25–100	>100	
Manganese ppm (Mn)		<100	100–500	>500	
Iron* ppm (Fe)		<75	75–250	>250	
Boron ppm (B)		<20	20–35	>35	
Molybdenum ppm (Mo)		<1	1–3	>3	

Telopea speciosissima
(Waratah)

Plant part Recently mature leaves — Ref: 17

Nutrient[1]	Deficient	Low	Normal	High	Excess
Nitrogen % (N)		<1.20	1.20–2.83		
Phosphorus % (P)			0.09–0.23	>0.23	
Potassium % (K)			1.00–1.75		
Calcium % (Ca)			0.53–1.10		
Magnesium % (Mg)			0.20–0.34		
Sulphur % (S)			0.23–0.35		
Sodium % (Na)			0.02–0.07		
Chloride % (Cl)			0.48–0.78		
Copper ppm (Cu)			<10		
Zinc ppm (Zn)			20–25		
Manganese ppm (Mn)			180–275		
Iron* ppm (Fe)			30–40		
Boron ppm (B)					
Molybdenum ppm (Mo)					

Thuja orientalis
(Chinese arborvitae)

Plant part 5.0–7.5 cm tips
Growth stage Young plants

Ref: 9

Nutrient[1]	Deficient	Low	Normal	High	Excess
Nitrogen % (N)			1.89		
Phosphorus % (P)			0.46		
Potassium % (K)			1.53		
Calcium % (Ca)			0.81		
Magnesium % (Mg)			0.39		
Sulphur % (S)					
Sodium % (Na)					
Chloride % (Cl)					
Copper ppm (Cu)					
Zinc ppm (Zn)			31		
Manganese ppm (Mn)			68		
Iron* ppm (Fe)			163		
Boron ppm (B)			27		
Molybdenum ppm (Mo)			1.33		

Tilia
(Lime tree or linden)

Sampling time Current year's terminals at mid season
Plant part Most recently mature leaf

Ref: 2

Nutrient[1]	Deficient	Low	Normal	High	Excess
Nitrogen % (N)			2.30–2.80		
Phosphorus % (P)			0.15–0.30		
Potassium % (K)			1.00–1.50		
Calcium % (Ca)			0.20–1.20		
Magnesium % (Mg)			0.15–0.30		
Sulphur % (S)					
Sodium % (Na)					
Chloride % (Cl)					
Copper ppm (Cu)			6–12		
Zinc ppm (Zn)			15–50		
Manganese ppm (Mn)			35–100		
Iron* ppm (Fe)					
Boron ppm (B)			15–40		
Molybdenum ppm (Mo)			0.05–0.20		

Veronica longifolia

Plant part Most recently mature leaf
Growth stage Flower buds visible

Ref: 1

Nutrient[1]	Deficient	Low	Normal	High	Excess
Nitrogen % (N)			2.9		
Phosphorus % (P)			0.35		
Potassium % (K)			1.20		
Calcium % (Ca)			0.58		
Magnesium % (Mg)			0.23		
Sulphur % (S)					
Sodium % (Na)					
Chloride % (Cl)					
Copper ppm (Cu)					
Zinc ppm (Zn)			57		
Manganese ppm (Mn)			30		
Iron* ppm (Fe)			82		
Boron ppm (B)			11		
Molybdenum ppm (Mo)					

Viburnum suspensum
(Viburnum)

Plant part Most recently mature leaf

Ref: 9

Nutrient[1]	Deficient	Low	Normal	High	Excess
Nitrogen % (N)		1.20–1.49	1.50–2.5	>2.5	
Phosphorus % (P)		0.10–0.14	0.15–0.4	>0.4	
Potassium % (K)		0.40–0.89	0.90–2.0	>2.0	
Calcium % (Ca)		0.35–0.59	0.60–1.5	>1.5	
Magnesium % (Mg)		0.20–0.24	0.25–1.0	>1.0	
Sulphur % (S)		0.13–0.19	0.20–0.4	>0.4	
Sodium % (Na)					
Chloride % (Cl)					
Copper ppm (Cu)		5–6	7–50	>50	
Zinc ppm (Zn)		16–19	20–200	>200	
Manganese ppm (Mn)		25–29	30–200	>200	
Iron* ppm (Fe)		25–29	30–200	>200	
Boron ppm (B)	<15	16–19	20–75	75–100	>101
Molybdenum ppm (Mo)		<1	1–3	>3	

Vinca

Plant part Most recently mature leaf — Ref: 6

Nutrient[1]	Deficient	Low	Normal	High	Excess
Nitrogen % (N)			4.9–5.4		
Phosphorus % (P)			0.4–0.6		
Potassium % (K)			2.9–3.6		
Calcium % (Ca)			1.4–1.6		
Magnesium % (Mg)			0.4–0.5		
Sulphur % (S)					
Sodium % (Na)					
Chloride % (Cl)					
Copper ppm (Cu)			5–10		
Zinc ppm (Zn)			40–45		
Manganese ppm (Mn)			165–300		
Iron* ppm (Fe)			95–150		
Boron ppm (B)			25–40		
Molybdenum ppm (Mo)					

Yucca elephantipes
(Yucca)

Plant part Most recently mature leaf — Ref: 7, 9

Nutrient[1]	Deficient	Low	Normal	High	Excess
Nitrogen % (N)		1.40–1.79	1.80–2.5	>2.5	
Phosphorus % (P)		0.10–0.14	0.15–0.8	>0.8	
Potassium % (K)		0.90–1.19	1.20–2.8	>2.8	
Calcium % (Ca)		0.70–0.99	1.00–2.5	>2.5	
Magnesium % (Mg)		0.15–0.19	0.20–0.6	>0.6	
Sulphur % (S)		0.12–0.19	0.20–0.8	>0.8	
Sodium % (Na)					
Chloride % (Cl)					
Copper ppm (Cu)		5–7	8–25	>25	
Zinc ppm (Zn)		16–19	20–200	>200	
Manganese ppm (Mn)		25–39	40–200	>200	
Iron* ppm (Fe)		20–24	25–200	>200	
Boron ppm (B)	<11	12–17	18–60	>60	
Molybdenum ppm (Mo)					

***Zantedeschia* spp.**
(Calla lily)

Plant part Most recently mature leaf
Growth stage Flower buds visible

Ref: 1

Nutrient[1]	Deficient	Low	Normal	High	Excess
Nitrogen % (N)			3.9		
Phosphorus % (P)			0.34		
Potassium % (K)			3.9		
Calcium % (Ca)			0.96		
Magnesium % (Mg)			0.23		
Sulphur % (S)					
Sodium % (Na)					
Chloride % (Cl)					
Copper ppm (Cu)					
Zinc ppm (Zn)			94		
Manganese ppm (Mn)			193		
Iron* ppm (Fe)			300		
Boron ppm (B)			13		
Molybdenum ppm (Mo)					

Zinnia elegans
(Zinnia)

Sampling time Prior to or at flowering
Plant part Most recently mature leaf

Ref: 2

Nutrient[1]	Deficient	Low	Normal	High	Excess
Nitrogen % (N)			2.00–5.00		
Phosphorus % (P)			0.20–0.45		
Potassium % (K)			2.20–5.50		
Calcium % (Ca)			0.80–2.00		
Magnesium % (Mg)			0.20–0.60		
Sulphur % (S)					
Sodium % (Na)					
Chloride % (Cl)					
Copper ppm (Cu)			4–10		
Zinc ppm (Zn)			25–70		
Manganese ppm (Mn)			40–120		
Iron* ppm (Fe)					
Boron ppm (B)			25–70		
Molybdenum ppm (Mo)			0.30–1.00		

References

1 Armitage, A.M. (1993). *Specialty Cut Flowers*. Variety Press/Timber Press, Oregon, USA.

2 Bergmann, W. (1992). *Nutritional Disorders of Plants: Development, Visual and Analytical Diagnosis*. Gustav Fischer Verlag, Jena, Stuttgart, New York.

3 Chase, A.R. and Broschat, T.K. (1991). *Diseases and Disorders of Ornamental Palms*. APS Press, Minnesota, USA.

4 Chase, A.R. (1989). Fertilising Calatheas with slow-release fertiliser. *Foliage Digest* **Oct.**, 1–3.

5 Cresswell, G.C. (1991). Assessing the phosphorus status of Proteas using plant analysis. *Proceedings 6th Biennial International Protea Association Conference Perth, September 22–27*. pp. 303–10.

6 Dole, J.M. and Wikins, H.F. (1989). Numbers from Guide values for nutrient element content of vegetables and flowers under glass. Glasshouse Crops Research Station, Aslsmeer and Naaldwijk, 1987. Assembled and updated by R.E. Widmer, University of Minnesota, June, 1985. *Grower Talks* **August 1989**.

7 Griffith, L.P. (1987). Boron nutrition in Florida ornamentals. *Proceedings Florida State Horticultural Society* **100**, 182–4.

8 Johanson, J. (1978). Effects of nutrient levels on growth, flowering and leaf nutrient content of greenhouse roses. *Acta Agriculturae Scandinavica* **28**, 1363–86.

9 Jones, B.J., Wolf, B. and Mills, H.A. (1991). *Plant Analysis Handbook 1. Methods of Plant Analysis and Interpretation*. Micro–Macro Publishing Inc., Athens, Georgia, USA.

10 Klossowski, W. and Strojny, Z. (1983). Chemical analysis of substrate and leaves as method of determining gerbera nitrogen and magnesium nutrition requirements. *Prace Instytutu Sadownictwa i Kwiaciarstwa w Skierniewicach* **8**, 111–21.

11 Lee, C.W., Pak, C.H., Choi, J.M. and Self, J.R. (1992). Induced micronutrient toxicity in *Petunia hybrida*. *Journal of Plant Nutrition* **15**(3), 327–39.

12 Lima, A.M.L.P. and Haag, H.P. (1981). Mineral nutrition of ornamental plants x. Deficiency symptoms of macronutrients, boron and iron in *Pelargonium zonale*. *Solo* **73**(1), 40–3.

13 Maier, N.A., Barth, G.E., Cecil, J.S., Chvyl, W.L. and Bartetzko, M.N. (1995). Effect of sampling time and leaf position on leaf composition of *Protea* 'Pink Ice'. *Australian Journal of Experimental Agriculture* **35**, 275–83.

14 Mastalerz, J.W. (1977). *The Greenhouse Environment*. John Wiley & Sons, New York. pp. 510–16.

15 McKay, M.E. (pers. comm.) (1976). Queensland Department of Primary Industries.

16 Nichols, D.G. and Beardsell, D.V. (1981). Interactions of calcium, nitrogen and potassium with phosphorus on the symptoms of toxicity in *Grevillea* cv. 'Poorinda Firebird'. *Plant and Soil* **61**, 437–45.

17 Parks, S., Cresswell, G.C. and Buining, F. (1996). Unpublished fertiliser trial data.

18 Parvin, P.E., Criley, R.A. and Bullock, R.M. (1973). Proteas: developmental research for a new cut flower crop. *Hort Science* **8**(4), 290–303.

19 Pool, R.T., Chase, A.R. and Conover, C.A. (1989). Chemical composition of good quality tropical plants. *Foliage Digest* **March**, 3.

20 Robinson, J.B., Barth, G.E. and Chvyl, L. (1992). Nutrition of container grown Sturt's desert pea – phosphorus. Horticultural Management, Adelaide.

21 Webb, M. and Pegrum, J. (1989). *Waxflower Nutrition Conference Proceedings. The Production and Marketing of Australian Flora*. University of Western Australia.

Appendix 2 Relative tolerance of some ornamental plants to saline growing conditions[1]

Very sensitive (threshold <2 dS/m)

Acanthus mollis	Bear's breeches
Begonia spp.	Begonia
Berberis thunbergii	Barberry
Camellia spp.	Camellia
Cedrus atlantica	Blue atlas cedar
Cotoneaster congestus	Pyrenees cotoneaster
Cotoneaster horizontalis	Rock cotoneaster
Cytisus × *praecox*	Broom
Dahlia spp.	Dahlia
Ensete ventricosum	Abyssinian banana
Euonymus alatus	Euonymus
Feijoa sellowiana	Pineapple guava
Fragaria chiloensis	Strawberry
Fuchsia spp.	Fuchsia
Gardenia spp.	Gardenia
Ilex cornuta	Chinese holly
Lilium spp.	Lily
Mahonia aquifolium	Oregon grape holly
Pachysandra terminalis	Japanese spurge
Persea americana	Avocado
Phormium tenax	New Zealand flax
Photinia × *fraseri* Robusta	Photinia
Picea pungens	Blue spruce
Podocarpus macrophyllus	Shrubby Japanese yew
Primula spp.	Primula
Pseudotsuga menziesii	Douglas fir
Rhododendron spp.	Azalea, rhododendron
Rosa spp.	Rose
Rubus spp.	Blackberry, boysenberry, etc.
Saintpaulia ionantha	African violet
Salix purpurea	Blue willow
Spiraea spp.	Spiraea
Tilia cordata	Linden
Trachelospermum jasminoides	Star jasmine
Viola hederaceae	Violet

[1] EC of a saturation extract from the growing medium. Table from *Growing Media for Ornamental Plants and Turf*, K. Handreck and N. Black, New South Wales University Press, 1984, with minor additions from *Agricultural Salinity Assessment and Management*, Ed. K.K. Tanji, American Society of Civil Engineers, 1990.

Sensitive (threshold <4 dS/m)

Abelia × grandiflora	Glossy abelia
Arbutus unedo	Irish strawberry tree
Aster spp.	Aster
Berberis × mentorensis	Barberry
Capsicum frutescens	Pepper
Cedrus deodara	Deodar cedar
Ceratonia siliqua	Carob
Cinnamomum camphora	Camphor tree
Clivia miniata	Kaffir lily
Crassula argentea	Jade plant
Euphorbia pulcherrima	Poinsettia
Felecia amelloides	Blue daisy
Ficus benjamina	Java fig
Forsythia × intermedia	Showy golden-bells
Gelsemium sempervirens	Carolina jasmine
Geranium spp.	Geranium
Gladiolus spp.	Gladiolus
Hedera canariensis	Algerian ivy
Hibiscus rosa-sinensis	Hibiscus
Ilex cornuta	Chinese holly cv. Burford
Juniperus virginiana	Eastern red cedar
Lagerstroemia indica	Crepe myrtle
Limonium perezii	Sea lavender
Liriodendron tulipifera	Tulip tree
Lonicera japonica	Honeysuckle
Magnolia grandiflora	Magnolia
Mathiola incana	Stock
Nandina domestica	Heavenly bamboo
Ophiopogon japonicus	Lily-turf
Pittosporum tobira	Pittosporum
Podocarpus macrophyllus	cv. Maki
Prunus amygdalus	Almond
P. domestica	Plum
P. persica	Peach
Pyracantha koidzumii	Firethorn
Pyrus armeniaca	Apricot
Rhamnus alternus	Italian blackthorn
Rosa sp.	Rose cv. Grenoble
Strelitzia reginae	Bird of paradise
Viburnum tinus	Viburnum
Vinea minor	Dwarf running myrtle
Washingtonia robusta	Cotton palm
Yucca filamentosa	Adam's-needle yucca
Zinnia elegans	Zinnia

Moderately tolerant (threshold <6 dS/m)

Agapanthus spp.	African lily
Agave attenuata	Century plant
Arecastrum romanzoffianum	Queen palm
Bauhinia purpurea	Orchid tree
Brassica oleracea cv. Capitata	Cabbage
Brunfelsia pauciflora	Yesterday, today and tomorrow
Buxus microphylla	Boxwood
Casuarina equisetifolia	Beach-oak
Chrysanthemum spp.	Chrysanthemum
Citrus limonea	Lemon
C. *paradisi*	Grapefruit
C. *sinensis*	Orange
Cupressus arizonica	Arizona cypress
C. *sempervirens*	Italian cypress
Cyclamen spp.	Cyclamen
Dianthus caryophyllus	Carnation
Dodonea viscosa	Dodonea
Elaeagnus pungens	–
Fraxinus pennsylvanica	Green ash
Hakea suaveolens	Sweet hakea
Juglans regia	Walnut
Juniperus chinensis	Spreading juniper
Lantana camara	Lantana
Ligustrum japonicum	Japanese privet
Magnolia grandiflora	Southern magnolia
Melaleuca leucadendron	Cajeput tree
Philodendron selloum	Philodendron
Phormium tenax	New Zealand flax
Pinus ponderosa	Ponderosa pine
P. thunbergiana	Japanese black pine
Pittosporum phillyraeoides	Desert willow
Platycladus orientalis	Oriental arborvitae
Prunus cerasifera	Cherry plum
Pyracantha graberi	Pyracantha
Pyrus communis	Pear
P. malus	Apple
Raphiolepis indica	Indian hawthorn
Salix vitellina	Golden willow
Shepherdia argentea	Buffaloberry
Syzygium paniculatum	Bush cherry
Thuja orientalis	Chinese arbovitae
Xylosma senticosa	Xylosma

Tolerant (threshold <8 dS/m)

Brahea edulis	Guadalupe palm
Callistemon viminalis	Bottlebrush
Calocephalus brownii	Pincushion bush
Casuarina glauca	Grey buloke
Chamaerops humilis	European fan palm
Cordyline indivisa	Blue dracaena
Crassula ovata	–
Dracaena endivisa	Dracaena
Elaeagnus angustifolia	Russian olive
Eucalyptus botryoides	Bangalay
E. coolabah	Coolibah
E. occidentalis	Flat-topped yate
E. pileata	Ravensthorpe mallee
E. robusta	Swamp mahogany
E. sideroxylon	Red ironbark
Euonymus japonica	Evergreen spindle-tree
Ficus carica	Fig
Gleditsia triacanthos	Honey locust
Hibiscus rosa-sinensis	Chinese hibiscus
Liquidambar styraciflua	Sweet gum
Melaleuca cuticularis	–
M. viminea	–
Metrosideros excelsa	NZ Christmas tree
Myoporum spp.	Boobialla
Nerium oleander	Oleander
Olea europaea	Olive
Ophiopogon jaburan	Lily-turf
Pinus halepensis	Allepo pine
Pittosporum crassifolium	Karo
Robinia pseudoacacia	Black locust
Rosmarinus officinalis	Rosemary
Vitis spp.	Grape

Very tolerant (threshold >8 dS/m)

Acacia cyanophylla	Orange wattle
A. *cyclops*	WA coastal wattle
A. *longifolia* var. *sophorae*	Coast wattle
A. *pulchella*	Western prickly moses
Alyxia buxifolia	Sea-box
Araucaria heterophylla	Norfolk Island pine
Arctotheca calendula	Capeweed
Atriplex spp.	Salt bush
Baccharis pilularis	Coyote bush
Banksia spp.	Banksia
Bougainvillea spectabilis	Bougainvillea
Callistemon citrinus	Crimson bottlebrush
Carissa grandiflora	Natal plum
Carpobrotus chilensis	Pigfaces
C. *edulis*	Pigfaces
Casuarina distyla	She oak
Coprosma repens	Mirror plant
Correa alba	White correa
Cortaderia sellowiana	Pampus grass
Delasperma spp.	Iceplants
Drosanthemum spp.	Iceplants, pigfaces
Eucalyptus camaldulensis	River red gum
E. *sargentii*	Salt river gum
E. *spathulata*	Swamp mallet
Ficus microcarpa	Small-leaf fig
Hibbertia scandens	Snake vine
Hymenocyclus spp.	Iceplants
Lagunaria patersonii	Norfolk Island hibiscus
Lampranthus spp.	Iceplants
Leptospermum laevigatum	Tea Tree
Leucophyllum frutescens	Texas sage
Lippia canescens repens	Lippia
Melaleuca armillaris	Bracelet honey-myrtle
M. *diosmifolia*	Cajeput tree
M. *nesophila*	Western tea-myrtle
Moraea vegeta	Iris
Pelargonium australe	Austral storks bill
Phoenix dactylifera	Date palm
Pyrus kawakamii	Evergreen pear
Platycladus orientalis	Oriental arbovita
Pinus pinea	Italian stone pine
Rhagodia spp.	Salt bush
Scaevola calendulacea	Dune fan flower
Tamarix aphylla	Tamarix
T. *pentandra*	Tamarix
Westringia fruticosa	Rosemary westringia
Yucca aloifolia	Spanish bayonet

Appendix 3 Relative sensitivity of plants to ozone[1]

Sensitive	Intermediate	Tolerant
Ash, green	Alder	Apricot
Ash, white	Apple, crab	Arbovitae
Aspen, quaking	Apricot, Chinese	Azalea, Chinese
Azalea, campfire	Begonia	Avocado
Azalea, hino	Boxelder	Beech, European
Azalea, Korean	Carnation	Birch, European white
Azalea, snow	Catalpa	Box, Japanese
Bridalwreath	Cedar, incense	Dogwood, grey
Browallia	Cherry, Lambert	Dogwood, white
Cherry, bing	Chrysanthemum	Euonymus, dwarf winged
Coleus	Elm, Chinese	Fir, balsam
Cotoneaster, rock	Fir, white	Fir, Douglas
Cotoneaster, speading	Fir, big-cone Douglas	Firethorn, Laland's fuchsia
Grape, Concord	Forsythia, cv. Lynwood Gold	Geranium
Lilac, Chinese	Grape, cv. Thompson Seedless	Gladiolus
Lilac, common	Gum, sweet	Gloxinia
Locust, honey	Honeysuckle, blue-leaf	Gum, black
Mountain ash, European	Larch, European	Hemlock, eastern
Oak, gambel	Larch, Japanese	Holly, American
Oak, white	Mock orange, sweet	Holly, Hetz's Japanese
Petunia	Oak, black	Impatiens
Pine, Austrian	Oak, pin	Ivy, English
Pine, Coulter	Oak, scarlet	Juniper, western
Pine, eastern white	Pine, knobcone	Lemon
Pine, Jack	Pine, lodgepole	Laurel, mountain
Pine, Jeffery	Pine, pitch	Linden, American
Pine, Monterey	Pine, Scotch	Linden, little-leaf
Pine, Ponderosa	Pine, sugar	Locust, black
Pine, Virginia	Pine, Torrey	Maple, Norway
Poplar, hybrid	Poinsettia	Maple, sugar
Poplar, tulip	Privet, common	Maple, red
Privet, londense	Redbud, eastern	Marigold
Snowberry	*Rhododendron catawbiense* cv. Album	Mimosa
Sumach	Rhododendron, cv. Nova Zembia	Oak, bur
Sycamore, American	Rhododendron, cv. Roseum Elegans	Oak, English
Tree-of-heaven	Silverberry	Oak, northern red
Walnut, English	Viburnum, linden	Oak, shingle
	Viburnum, tea	Pachysandra
	Willow, weeping	Pagoda, Japanese
		Peach
		Pear, Bartlett
		Periwinkle
		Pieris, Japanese
		Pine, digger
		Pine, singleleaf pinyon
		Pine, red
		Privet, Amur north
		Redwood
		Rhododendron, cv. Carolina
		Sequoia, giant
		Snapdragon
		Spiraea
		Spruce, Black Hills
		Spruce, Colorado blue
		Spruce, Norway
		Spruce, white
		Viburnum, Koreanspice
		Viburnum
		Virginia-creeper
		Walnut, black
		Yew, dense
		Yew, Hatfield's pyramidal
		Zinnia

[1] From USEPA (1978). Diagnosing vegetation injury caused by air pollution. USEPA 450/3-78-005.

Appendix 4 Relative sensitivity of plants to sulphur dioxide[1]

Sensitive	Tolerant
Alder	Cedar
Apple	Citrus
Aster	Linden
Austrian pine	Maple
Bachelor button	
Cosmos	
Douglas fir	
Four o'clock	
Gladiolus	
Hazel	
Jack pine	
Larch	
Large-toothed aspen	
Ponderosa pine	
Red pine	
Sweet William	
Sweet pea	
Trembling aspen	
Tulip	
Verbena	
Violet	
White pine	
White birch	
Willow	
Zinnia	

[1] From USEPA (1978). Diagnosing vegetation injury caused by air pollution. USEPA 450/3-78-005.

Appendix 5 Relative sensitivity of ornamental plants to fluoride in the air[1]

High sensitivity	Medium sensitivity	Low sensitivity
Flowers		
Aster	Azalea	Ageratum
Begonia	Dahlia	Alyssum
Carnation	Geranium	Amaryllis
Day lily	Gypsophila	Bridal wreath
Gerbera	Impatiens	Calendula
Gladiolus	Lilac	Camellia
Iris	Narcissus	Chrysanthemum
Poinsettia	Peony	Cattleya orchid
Tulip	Petunia	Dendrobium
	Rhododendron	Dusty miller
	Rose	Firethorn
	Snapdragon	Gazania
	Sweet William	Gloxinia
	Violet	Marigold
		Phalaenopsis
		Salvia
Foliage		
Asparagus sprengeri	Maranta	Aglaonema
Chlorophytum	Rhoeo	Boston fern
Cordyline		Bromeliad
Dracaena warneckii		Caladium
Philodendron oxycardium		Crassula
Philodendron panduriforme		Dieffenbachia
Pteris		*Dracaena massangeana*
		Fluffy ruffle
		Gynura
		Hemigraphis
		Hoya
		Nephthytis
		Peperomia
		Philodendron cv. Red Emerald
		Pilea
		Plectanthrus
		Pothos
		Privet
		Sanseveria
		Tradescantia

[1] Compiled using tables from Woltz, S.S. and Waters, W.E. (1976) Susceptibility of some ornamentals to fluoride air pollution, Bradenton AREC Res. Rept, G.C. 1976–77, and USEPA (1978) Diagnosing vegetation injury caused by air pollution, USEPA 450/3-78-005.

Appendix 6 Relative sensitivity of plants to nitrogen dioxide[1]

Sensitive	Intermediate	Tolerant
Trees and shrubs		
European larch	Fir	Black locust
European white birch	Japanese maple	Elder
Japanese larch	Linden	Elm
	Norway maple	European hornbeam
	Spruce	Gingko
		Oak
		Pine
		Purple beech
		Yew
Flowering plants		
Azalea	Brittlewood	Carissa
Bougainvillea	Cape jasmine	Croton
Hibiscus	Catawba rhododendron	Dahlia
Oleander	Fuchsia	Daisy
Pyracantha	Gardenia	Gladiolus
Rose	Ixora	Heath
Snapdragon	Ligustrum	Lily of the valley
Sweet pea	Petunia	Plantain lily
Tuberous begonia	Pittosporum	Shore juniper

[1] From USEPA (1978) Diagnosing vegetation injury caused by air pollution, USEPA 450/3-78-005.

Appendix 7 Relative sensitivity of plants to ethylene[1]

Sensitive	Intermediate	Tolerant
African marigold	Ageratum	Acacia
Orchid	Carnation	Calendula
	Larkspur	China aster
	Lily	Dahlia
	Rose	Forget-me-not
	Snapdragon	Lobelia
	Zinnia	Sweet pea
		Viola

[1] Arbitrary classification based on available data (USEPA, 1978).